TEXTS AND READINGS
IN MATHEMATICS

7

Harmonic Analysis
Second Edition

Texts and Readings in Mathematics

Harmonic Analysis
Second Edition

Henry Helson

HINDUSTAN
BOOK AGENCY

Published by
Hindustan Book Agency (India)
P 19 Green Park Extension
New Delhi 110 016
India

email: info@hindbook.com
http://www.hindbook.com

ISBN 978-93-80250-05-2

CONTENTS

1. **Fourier Series and Integrals**
   1.1 Definitions and easy results 1
   1.2 The Fourier transform 7
   1.3 Convolution, approximate identities, Fejér's theorem 11
   1.4 Unicity theorem, Parseval relation; Fourier-
       Stieltjes coefficients 17
   1.5 The classical kernels 25
   1.6 Summability: metric theorems 30
   1.7 Pointwise summability 35
   1.8 Positive definite sequences; Herglotz' theorem ... 40
   1.9 The inequality of Hausdorff and Young 42
   1.10 Measures with bounded powers; endomorphisms of l^1 45

2. **The Fourier Integral**
   2.1 Introduction 53
   2.2 Kernels on **R** 56
   2.3 The Plancherel theorem 62
   2.4 Another convergence theorem;
       the Poisson summation formula 65
   2.5 Bochner's theorem 69
   2.6 The continuity theorem 74

3. **Discrete and Compact Groups**
   3.1 Characters of discrete groups 79
   3.2 Characters of compact groups 87
   3.3 Bochner's theorem 90
   3.4 Examples 93
   3.5 Minkowski's theorem 97
   3.6 Measures on infinite product spaces 100
   3.7 Continuity of seminorms 101

4. **Hardy Spaces**
   4.1 $H^p(\mathbf{T})$ 105
   4.2 Invariant subspaces; factoring;
       proof of the theorem of F. and M. Riesz ... 110
   4.3 Theorems of Szegő and Beurling 118
   4.4 Structure of inner functions 124
   4.5 Theorem of Hardy and Littlewood;
       Hilbert's inequality 129
   4.6 Hardy spaces on the line 134

5. Conjugate Functions

5.1 Conjugate series and functions   143
5.2 Theorems of Kolmogorov and Zygmund   146
5.3 Theorems of Riesz and Zygmund   152
5.4 The conjugate function as a singular integral   ...   157
5.5 The Hilbert transform   163
5.6 Maximal functions   165
5.7 Rademacher functions; absolute Fourier multipliers   170

6. Translation

6.1 Theorems of Wiener and Beurling;
     the Titchmarsh convolution theorem   181
6.2 The Tauberian theorem   185
6.3 Spectral sets of bounded functions   191
6.4 A theorem of Szegő; the theorem of Grużewska
     and Rajchman; idempotent measures   199

7. Distribution

7.1 Equidistribution of sequences   205
7.2 Distribution of $(n_k u)$ 209
7.3 $(k^r u)$ 211

Appendix
Integration by parts   219

Bibliographic Notes   221

Index   225

PREFACE TO THE SECOND EDITION

Harmonic Analysis used to go by the more prosaic name Fourier Series. Its elevation in status may be due to recognition of its crucial place in the intersection of function theory, functional analysis, and real variable theory; or perhaps merely to the greater weightiness of our times. The 1950's were a decade of progress, in which the author was fortunate to be a participant. Some of the results from that time are included here.

This book begins at the beginning, and is intended as an introduction for students who have some knowledge of complex variables, measure theory, and linear spaces. Classically the subject is related to complex function theory. We follow that tradition rather than the modern direction, which prefers real methods in order to generalize some of the results to higher dimensional spaces. In this edition there is a full presentation of Bochner's theorem, and a new chapter treats the duality theory for compact and discrete abelian groups. Then the author indulges his own experience and tastes, presenting some of his own theorems, a proof by C. L. Siegel of Minkowski's theorem, applications to probability, and in the last chapter two different methods of proving the theorem of Weyl on equidistribution modulo 1 of $(P(k))$, where P is a real polynomial with at least one irrational coefficient.

This is not a treatise. If what follows is interesting and useful, no apology is offered for what is not here. The notes at the end are intended to orient the reader who wishes to explore further.

I express my warm thanks to Robert Burckel, whose expert criticism and suggestions have been most valuable. I am also indebted to Alex Gottlieb for detailed reading of the text. This edition is appearing simultaneously in India. I am grateful to Professor R. Bhatia, and to the Hindustan Book Agency for the opportunity to present it in this cooperative way.

HH

Chapter 1
Fourier Series and Integrals

1. Definitions and easy results

The unit circle $\mathbf{T}$ consists of all complex numbers of modulus 1. It is a compact abelian group under multiplication. If f is a function on $\mathbf{T}$, we can define a periodic function F on the real line $\mathbf{R}$ by setting $F(x) = f(e^{ix})$. It does not matter whether we study functions on $\mathbf{T}$ or periodic functions on $\mathbf{R}$; generally we shall write functions on $\mathbf{T}$. Everyone knows that in this subject the factor 2π appears constantly. Most of these factors can be avoided if we replace Lebesgue measure dx on the interval $(0, 2\pi)$ by $d\sigma(x) = dx/2\pi$. We shall generally omit the limits of integration when the measure is σ; they are always 0 and 2π, or another interval of the same length.

One more definition will simplify formulas: χ is the function on $\mathbf{T}$ with values $\chi(e^{ix}) = e^{ix}$. Thus χ^n represents the exponential e^{nix} for each integer n.

We construct Lebesgue spaces $\mathbf{L}^p(\mathbf{T})$ with respect to σ, $1 \le p \le \infty$. The spaces $\mathbf{L}^1(\mathbf{T})$ of summable functions and $\mathbf{L}^2(\mathbf{T})$ of square-summable functions are of most interest. Since the measure is finite these spaces are nested: $\mathbf{L}^p(\mathbf{T}) \supset \mathbf{L}^r(\mathbf{T})$ if $p < r$ (Problem 2 below). Thus $\mathbf{L}^1(\mathbf{T})$ contains all the others. For summable functions f we define *Fourier coefficients*

$$(1.1) \qquad a_n(f) = \int f\chi^{-n}\, d\sigma \quad (n = 0, \pm 1, \pm 2, \ldots),$$

and then the *Fourier series* of f is

$$(1.2) \qquad f(e^{ix}) \sim \sum a_n(f) \, e^{nix}.$$

We do not write equality in (1.2) unless the series converges to f.

There is a class of functions for which this is obviously the case. A *trigonometric polynomial* is a finite sum

$$(1.3) \qquad P(e^{ix}) = \sum a_n \, e^{nix}.$$

Then

$$(1.4) \qquad a_n(P) = \sum_k a_k \int e^{(k-n)ix} \, d\sigma(x) = a_n,$$

if we define $a_n = 0$ for values of n not occurring in the sum (1.3). Thus (1.3), which defines P, is also the Fourier series of P.

This reasoning can be carried further. Suppose that f is a function defined as the sum of the series in (1.3), now allowed to have infinitely many terms but assumed to converge *uniformly* on **T** for some ordering of the series. Then the same calculation is valid and we find that $a_n(f) = a_n$ for each n. That is, the trigonometric series converging to f is also the Fourier series of f.

From (1.1) we obviously have $|a_n(f)| \leq \|f\|_1$ for all n. A more precise result can be proved in $\mathbf{L}^2(\mathbf{T})$.

Bessel's Inequality. *If f is in* $\mathbf{L}^2(\mathbf{T})$, *then*

$$(1.5) \qquad \sum |a_n(f)|^2 \leq \|f\|_2^2.$$

Part of the assertion is that the series on the left converges. For each positive integer k set

(1.6)
$$f_k = \sum_{-k}^{k} a_n(f)\chi^n.$$

The norm of any function is non-negative, and thus

(1.7)
$$0 \le \|f - f_k\|_2^2 = \|f\|_2^2 + \|f_k\|_2^2 - 2\Re \int f \bar{f}_k \, d\sigma.$$

Since the exponentials form an orthonormal system, the second term on the right equals

(1.8)
$$\sum_{-k}^{k} |a_n|^2.$$

The last term is

(1.9)
$$-2\Re \sum_{-k}^{k} \bar{a}_n \int f \chi^{-n} \, d\sigma = -2 \sum_{-k}^{k} |a_n|^2.$$

This term combines with (1.8), and (1.7) becomes

(1.10)
$$\sum_{-k}^{k} |a_n|^2 \le \|f\|^2.$$

Since k is arbitrary, (1.5) is proved.

A kind of converse to Bessel's inequality is the

Riesz-Fischer theorem. *If (a_n) is any square-summable sequence, there is a function f in $\mathbf{L}^2(\mathbf{T})$ such that $a_n(f) = a_n$ for all n, and*

(1.11)
$$\sum |a_n|^2 = \|f\|_2^2.$$

Define

$$(1.12) \qquad f_k = \sum_{-k}^{k} a_n \chi^n$$

for each positive integer N. Then

$$(1.13) \qquad \|f_{k+r} - f_k\|_2^2 = \sum_{k < |n| \le k+r} |a_n|^2$$

for all positive integers r. Thus (f_k) is a Cauchy sequence in $\mathbf{L}^2(\mathbf{T})$. Let f be its limit. The sequence $(a_n(f))$ converges to $(a_n(f))$ in the space $\mathbf{l}^2$, by Bessel's inequality. Therefore (1.11), which holds for each k, is valid in the limit.

Now Bessel's inequality is actually *equality* for every f in $\mathbf{L}^2(\mathbf{T})$, and this equality is called the *Parseval relation.* Computations already performed show that equality holds for all trigonometric polynomials. The Fourier transform, thought of as a mapping from trigonometric polynomials in the norm of $\mathbf{L}^2(\mathbf{T})$ into $\mathbf{l}^2$, is an isometry whose range consists of all sequences (a_n) such that $a_n = 0$ for $|n|$ sufficiently large. The range is dense in $\mathbf{l}^2$. If we knew that the family of trigonometric polynomials is dense in $\mathbf{L}^2(\mathbf{T})$, then the isometry has a unique continuous extension to a linear isometry of all of $\mathbf{L}^2(\mathbf{T})$ onto $\mathbf{l}^2$. It is obvious that this extension is the Fourier transform. In Section 4 it will be shown that trigonometric polynomials are indeed dense in $\mathbf{L}^2(\mathbf{T})$, and this will prove the Parseval relation.

When the Parseval relation is known, the Riesz-Fischer theorem can immediately be strengthened to say that *the function whose coefficients are the given sequence (a_n) is unique.* For if f and g have the same Fourier coefficients, then $f - g$ has all its coefficients 0; that is, $f - g$ is orthogonal in the Hilbert space to all trigonometric polynomials, and must be null.

Mercer's theorem. *For all f in* $\mathbf{L}^1(\mathbf{T})$, $a_n(f)$ *tends to 0 as n tends to* $\pm\infty$.

Bessel's inequality shows that this is true if f is in $\mathbf{L}^2(\mathbf{T})$. Now $\mathbf{L}^2(\mathbf{T})$ is dense in $\mathbf{L}^1(\mathbf{T})$. (A proof of this fact depends on the particular way in which measure theory was developed and the Lebesgue spaces defined.) Choose a sequence (f_k) of elements of $\mathbf{L}^2(\mathbf{T})$ converging to f in the norm of $\mathbf{L}^1(\mathbf{T})$. Then $(a_n(f_k))$ converges to $(a_n(f))$ as k tends to ∞, *uniformly in n*. It follows that the limit sequence vanishes at $\pm\infty$, as was to be proved.

Mercer's theorem is the source of theorems asserting the convergence of Fourier series. Here is the most important such result, with a simple proof suggested by Paul Chernoff.

Theorem 1. *Suppose that f is in* $\mathbf{L}^1(\mathbf{T})$ *and that* $f(e^{ix})/x$ *is summable on* $(-\pi, \pi)$. *Then*

$$(1.14) \qquad \sum_{-M}^{N} a_n(f) \to 0$$

as M, N tend independently to ∞.

Form the function $g(e^{ix}) = f(e^{2ix})/\sin x$. The hypothesis implies that g is in $\mathbf{L}^1(\mathbf{T})$. Let (a_n) be the Fourier coefficients of f, and (b_n) those of g. We have

$$(1.15) \qquad f(e^{2ix}) = \frac{1}{2i}(e^{ix} - e^{-ix})\, g(e^{ix});$$

calculating the coefficients with even indices of the functions on both sides gives

$$(1.16) \qquad 2ia_n = b_{2n-1} - b_{2n+1}$$

for each integer n. Hence

$$(1.17) \qquad 2i \sum_{-M}^{N} a_n = b_{-2M-1} - b_{2N+1}.$$

By Mercer's theorem, this quantity tends to 0 as M, N tend to ∞.

   Corollary. *If f is in $\mathbf{L}^1(\mathbf{T})$ and satisfies a Lipschitz condition at a point e^{it}, then the Fourier series of f converges to $f(e^{it})$ at that point.*

   It is easy to check that addition of a constant to a summable function, and translation, have the formally obvious effect on the Fourier series of the function (Problem 1 below). Therefore without loss of generality we may assume that $t = 0$ and $f(1) = 0$. Now the hypothesis means that

$$(1.18) \qquad |f(e^{ix}) - f(e^{it})| \le k|x - t|^{\alpha}$$

for some constant k, and some α satisfying $0 < \alpha \le 1$, for all x. With f and t as just assumed, the hypothesis of the theorem is satisfied. The conclusion is that f is the sum of its Fourier series at the point 1, and the general result follows.

   The results of this section are striking but elementary. In order to get further we shall have to introduce new techniques.

Problems

1. Show that if f is in $\mathbf{L}^1(\mathbf{T})$, and g is defined by

$$g(e^{ix}) = c + f(e^{i(x+s)})$$

where c is complex and s real, then $a_n(g) = a_n(f)\, e^{nis}$ for all $n \ne 0$, and $a_0(g) = a_0(f) + c$.

2. Suppose that $1 \le p < r \le \infty$. Use Hölder's inequality to show that (a) $L^p(\mathbf{T}) \supset L^r(\mathbf{T})$ and (b) $||f||_p \le ||f||_r$ for f in $L^r(\mathbf{T})$.

3. Show that if f is real, then its coefficients satisfy $a_{-n} = \bar{a}_n$ for all n. (In particular, a_0 is real.)

4. Calculate the Fourier series of these periodic functions.

    (a) $f(e^{ix}) = -1$ on $(-\pi, 0)$, $= 1$ on $(0, \pi)$

    (b) $g(e^{ix}) = x + \pi$ on $(-\pi, 0)$, $= x - \pi$ on $(0, \pi)$

    (c) $h(e^{ix}) = (1 - re^{ix})^{-1}$, where $0 < r < 1$

(These series will be needed later; keep a record of them.)

5. Prove the *principle of localization*: *if f and g are in* $L^1(\mathbf{T})$ *and are equal on some interval, then at each interior point of the interval their Fourier series both converge and to the same value, or else both diverge.*

6. Suppose that $f(e^{ix})$ and $(f(e^{ix}) + f(e^{-ix}))/x$ are summable on $(-\pi, \pi)$. Show that

$$\sum_{-N}^{N} a_n(f) \to 0$$

as $N \to \infty$. Is the conclusion of Theorem 1 necessarily true?

2. The Fourier transform

On the line $\mathbf{R}$ construct the Lebesgue spaces $L^p(\mathbf{R})$ with respect to ordinary Lebesgue measure. The Fourier transform of a summable function f is

$$(2.1) \qquad \hat{f}(y) = \int_{-\infty}^{\infty} f(x) \, e^{-ixy} \, dx.$$

Obviously $|\hat{f}(y)| \le ||f||_1$ for all y. The function $\hat{f}$ is continuous. To see this, write

$$(2.2) \qquad \hat{f}(y') - \hat{f}(y) = \int_{-\infty}^{\infty} f(x) \, (e^{-ixy'} - e^{-ixy}) \, dx.$$

As y' tends to y the integrand tends to 0 for each x. Also the modulus of the integrand does not exceed $2|f(x)|$, a summable function. By Lebesgue's dominated convergence theorem, the integral tends to 0 as claimed.

The analogue of Mercer's theorem is called the *Riemann-Lebesgue lemma: $\hat{f}(y)$ tends to 0 as y tends to $\pm\infty$.* If a sequence of functions f_n converges to f in $\mathbf{L}^1(\mathbf{R})$, then the transforms $\hat{f}_n$ tend to $\hat{f}$ uniformly. Therefore, as in the proof of Mercer's theorem, it will suffice to show that the assertion is true for all functions in a dense subset of $\mathbf{L}^1(\mathbf{R})$. Let f be the characteristic function of an interval $[a, b]$. Its transform is

$$(2.3) \qquad \int_a^b e^{-ixy}\, dx = \frac{e^{-iay} - e^{-iby}}{iy} \qquad (y \neq 0),$$

which tends to 0. Therefore a linear combination of characteristic fur 'ions of intervals, that is a step function, has the property, ə ₁ such functions are dense. This completes the proof.

There is a version of Theorem 1 for the line.

Theorem 1'. *If f and $f(x)/x$ are summable, then*

$$(2.4) \qquad \lim_{A,\,B\to\infty} \int_{-A}^{B} \hat{f}(y)\, dy = 0.$$

The quantity that should tend to 0 as A, B tend to ∞ is

$$(2.5) \qquad \int_{-A}^{B} \int_{-\infty}^{\infty} f(x)\, e^{-ixy}\, dx\, dy.$$

The integrand is summable over the product space, so it is legitimate to interchange the order of integration. After integrating with respect to y we find

$$(2.6) \qquad \int_{-\infty}^{\infty} f(x)\,(-ix)^{-1}\big(e^{-iBx} - e^{iAx}\big)\,dx.$$

This tends to 0 by the Riemann-Lebesgue lemma.

There is an inversion theorem like the Corollary to Theorem 1, but a difficulty has to be met that did not arise on the circle.

Corollary. *If f is summable on the line and satisfies a Lipschitz condition at t, then*

$$(2.7) \qquad f(t) = \lim_{A,\,B \to \infty} \frac{1}{2\pi} \cdot \int_{-A}^{B} \hat{f}(y)\,e^{ity}\,dy.$$

As on the circle, we may assume that $t = 0$. If $f(0) = 0$ there is nothing more to prove. However, we cannot now reduce the general result to this one merely by subtracting $f(0)$ from f, because nonzero constants are not summable functions. Therefore we must subtract from f a *function* g that is summable, smooth near 0, with $g(0) = f(0)$, such that we can actually calculate $\hat{g}$ and the inverse Fourier transform of $\hat{g}$. The details are in Problem 5 below.

These theorems are parallel to the ones proved on the circle group in the last section. Analysts have known for a hundred years that this similarity goes very far, although proofs on the line are often more complicated. However our axiomatic age has produced a unified theory of harmonic analysis on locally compact abelian groups, among which are the circle, the line and the integer group, but also other groups of interest in analysis and number theory. The discovery of Banach algebras by A. Beurling and I. M. Gelfand was closely associated with this generalization of classical harmonic analysis. In beginning the study of Fourier series it is well to realize that its ideas are more general than appears in the classical context. Furthermore, this commutative theory has inspired much of the theory of group representations, which is the generalization of harmonic analysis to locally compact, non-abelian groups.

There is one more classical Fourier transform. The transform of a function on **T** is a sequence, that is, a function on **Z**. A function on **R** has transform that is another function on **R**. Now let (a_n) be a summable function on **Z**. Its transform is the function on **T** defined by

$$(2.8) \qquad f(e^{ix}) = \sum_{-\infty}^{\infty} a_n e^{-nix}.$$

Generally, the transform of a function on a locally compact abelian group is a continuous function on the dual of that group. The dual of **T** is **Z**, the dual of **Z** is **T**, and **R** is dual to itself.

Problems

1. Calculate the Fourier transforms of (a) the characteristic function of the interval $[-A, A]$; (b) the triangular function vanishing outside $(-A, A)$, equal to 1 at the origin, and linear on the intervals $(-A, 0)$ and $(0, A)$.

2. Express the Fourier transform of $f(x+t)$ (a function of x) in terms of $\hat{f}$.

3. Suppose that $(1 + |x|)f(x)$ is summable. Show that $(\hat{f})'$ is the Fourier transform of $-ixf(x)$, as suggested by differentiating (2.1).

4. Find the Fourier transform of f' if f is continuously differentiable with compact support. (Such functions are dense in

$\mathbf{L^1(R)}$; this gives a new proof of the Riemann-Lebesgue lemma.)

5. Calculate the Fourier transform of $g(x) = e^{-|x|}$. Verify that

$$\frac{1}{2\pi} \int_{-\infty}^{\infty} \hat{g}(y) \, dy = 1.$$

Use this information to complete the proof of the corollary.

6. (a) Use the calculus of residues to find

$$\frac{1}{2\pi} \int_{-\infty}^{\infty} \frac{2}{1 + y^2} e^{ixy} \, dy.$$

(b) Obtain the same result by applying the inversion theorem to the function of Problem 5.

7. Show that the Fourier transform of $\exp(-x^2)$ is $\sqrt{\pi} \exp(-y^2/4)$. [Find the real integral

$$\int_{-\infty}^{\infty} \exp(-x^2 + 2xu) \, dx.$$

Show that this is an entire function of u, and complexify u.]

3. Convolution; approximate identities; Fejér's theorem

The *convolution* of functions in $\mathbf{L^1(T)}$ is defined by

$$(3.1) \qquad f * g(e^{ix}) = \int f(e^{it}) \, g(e^{i(x-t)}) \, d\sigma(t).$$

If f and g are square-summable the integral exists, and is bounded by $\|f\|_2 \|g\|_2$ by the Schwarz inequality. More generally, if f is in $\mathbf{L^p(T)}$ and g is in $\mathbf{L^q(T)}$, where p and q are conjugate exponents, then the integral exists, and $|f * g(e^{ix})| \le \|f\|_p \|g\|_q$ for all x, by the Hölder inequality. Furthermore the convolution is a continuous function (Problem 4 below).

If f and g are merely assumed to be summable, their

product may not be summable, and the integral may not exist. It is surprising and important that nevertheless the product under the integral sign in (3.1) is summable at *almost every* point. To prove this form the double integral

$$(3.2) \qquad \int\int |f(e^{it})\, g(e^{i(x-t)})|\, d\sigma(t)\, d\sigma(x).$$

The integrand is non-negative and measurable on the product space, so the integral exists, finite or infinite. By Fubini's theorem, it equals either of the two iterated integrals. Integrating first with respect to x, and using the fact that σ is invariant under translation, we find $||f||_1 ||g||_1$ as the value of (3.2). Therefore the integral with respect to t

$$(3.3) \qquad h(e^{ix}) = \int |f(e^{it})\, g(e^{i(x-t)})|\, d\sigma(t)$$

must be finite a.e. Moreover h is summable, and $||h||_1 \le ||f||_1 ||g||_1$. It follows that the integrand in (3.1) is summable for a.e. x, the convolution (defined almost everywhere) belongs to $\mathbf{L}^1(\mathbf{T})$, and

$$(3.4) \qquad ||f*g||_1 \le ||f||_1 ||g||_1.$$

Convolution is associative and commutative, and distributes over addition. (The proofs are calculations involving elementary changes of variable, and are asked for in Problem 1 below.) Thus $\mathbf{L}^1(\mathbf{T})$ is a *commutative Banach algebra*, and indeed this is the algebra that led Gelfand to the concept. This algebra has no identity.

(An *algebra* is a ring admitting multiplication by scalars

from a specified field. Most rings in analysis are algebras over the real or complex numbers. An *ideal* in an algebra is an ideal (in the sense of rings) that is invariant under multiplication by scalars.)

Convolution is defined analogously on **R** and on **Z**. On **R**

$$(3.5) \qquad f*g(x) \;=\; \int_{-\infty}^{\infty} f(t)\,g(x-t)\,dt.$$

Once more the integral is absolutely convergent if f and g are in $\mathbf{L}^2(\mathbf{R})$, or if f and g belong to complementary Lebesgue spaces, and the convolution is continuous. If the functions are merely summable, the Fubini theorem shows, exactly as on the circle, that the integrand is summable for almost every x, $f*g$ is summable, and (3.4) holds.

If f and g are in $\mathbf{l}^1$, the definition is

$$(3.6) \qquad f*g(n) \;=\; \sum_{m=-\infty}^{\infty} f(m)\,g(n-m).$$

This series converges absolutely for all n, and (3.4) holds once more.

An *approximate identity* on **T** is a sequence of functions (e_n) with these properties: each e_n is non-negative, has integral 1 with respect to σ, and for every positive number $\epsilon < \pi$

$$(3.7) \qquad \lim_{n\to\infty} \int_{-\epsilon}^{\epsilon} e_n\,d\sigma \;=\; 1.$$

This means that nearly all the area under the graph of e_n is close to the origin if n is large. Although $\mathbf{L}^1(\mathbf{T})$ has no identity under convolution, it has approximate identities, and these almost serve the same purpose, by this fundamental result:

Fejér's Theorem. *Let f belong to* $\mathbf{L}^p(\mathbf{T})$ *with* $1 \le p < \infty$. *For any approximate identity* (e_n), $e_n * f$ *converges to f in the norm of* $\mathbf{L}^p(\mathbf{T})$. *If f is in* $\mathbf{C}(\mathbf{T})$, *the space of continuous functions on* $\mathbf{T}$, *then* $e_n * f$ *converges to f uniformly.*

We prove the second assertion first. Let f be continuous on $\mathbf{T}$. Then

$$(3.8) \qquad f(e^{ix}) - e_n * f(e^{ix}) = \int [f(e^{ix}) - f(e^{i(x-t)})]\, e_n(e^{it})\, d\sigma(t),$$

because e_n has integral 1. Since f is continuous and $\mathbf{T}$ compact, f is uniformly continuous: given any positive number ϵ, there is a positive δ such that $|f(e^{ix}) - f(e^{i(x-t)})| \le \epsilon$ for all x, provided that $|t| \le \delta$. Denote by I the part of the integral over $(-\delta, \delta)$, and by J the integral over the complementary subset of $\mathbf{T}$. For all n

$$(3.9) \qquad\qquad |I| \le \epsilon \int e_n\, d\sigma = \epsilon.$$

If M is an upper bound for $|f|$, we have

$$(3.10) \qquad\qquad |J| \le 2M \int_{\delta}^{2\pi-\delta} e_n\, d\sigma,$$

and this quantity is as small as we please provided that n is large enough. These estimates do not depend on x. Therefore the difference (3.8) is uniformly as small as we please if n is large enough, which is what was to be proved.

Now let p satisfy $1 \le p < \infty$. For continuous f, the uniform convergence of $e_n * f$ to f implies convergence in $\mathbf{L}^p(\mathbf{T})$. In order to extend this result from continuous functions to all functions in

$\mathbf{L}^p(\mathbf{T})$, we use the following general principle, which has many applications besides this one.

Principle. *Let* $\mathbf{X}$ *and* $\mathbf{Y}$ *be normed vector spaces, and* $\mathbf{Y}$ *a Banach space. Let* (T_n) *be a sequence of linear operators from* $\mathbf{X}$ *to* $\mathbf{Y}$ *whose bounds* $||T_n||$ *are all less than a number* K. *Suppose that* $T_n x$ *converges to a limit we call* Tx, *for each* x *in a dense subset of* $\mathbf{X}$. *Then* $T_n x$ *converges for all* x *in* $\mathbf{X}$, *and the limit* Tx *defines a linear operator* T *with bound at most* K.

The easy proof is omitted.

Lemma. *For* f *in* $\mathbf{L}^1(\mathbf{T})$ *and* g *in* $\mathbf{L}^p(\mathbf{T})$, $1 \le p$, $f * g$ *is in* $\mathbf{L}^p(\mathbf{T})$ *and* $||f*g||_p \le ||f||_1 ||g||_p$.

The lemma is trivial if $p = \infty$, so we assume p finite. Let q be the conjugate exponent, and h any function in $\mathbf{L}^q(\mathbf{T})$ with norm 1. As a linear functional on $f*g$, h has the value

$$(3.11) \qquad h*(f*g)(1) = \int \int h(e^{-ix}) f(e^{it}) g(e^{i(x-t)}) \, d\sigma(t) \, d\sigma(x).$$

The double integral exists absolutely by Hölder's inequality, as we see by integrating first with respect to x, and in modulus does not exceed $||f||_1 ||g||_p$ (because h has norm 1 in $\mathbf{L}^q(\mathbf{T})$). This proves that $f*g$ belongs to $\mathbf{L}^p(\mathbf{T})$ with norm at most equal to this quantity, and the lemma is proved.

Now we can finish the proof of Fejér's theorem. Convolution with e_n is a linear operation in $\mathbf{L}^p(\mathbf{T})$ with bound at most 1, by the lemma. This sequence of operators converges to the identity operator on each element of $\mathbf{C}(\mathbf{T})$, a dense subset of $\mathbf{L}^p(\mathbf{T})$. (This result comes from integration theory; the point is discussed further below.) The Principle asserts that the sequence of operators converges everywhere, and the limit is a bounded operator.

Hence this limit is the identity operator on $\mathbf{L}^p(\mathbf{T})$. This proves the theorem.

The lemma is an instance of a general fact. Let S_t be a linear operator in a Banach space $\mathbf{X}$ for each real number t. For any g in $\mathbf{X}$, $(S_t g)$ is a mapping from real numbers to $\mathbf{X}$. For f in $\mathbf{L}^1(\mathbf{T})$, the generalized convolution

$$(3.12) \qquad f * g = \int f(e^{it}) S_t g \, d\sigma$$

may have a meaning as a limit of sums of elements of $\mathbf{X}$. In our case, S_t is the translation operator

$$(3.13) \qquad (S_t g)(e^{ix}) = g(e^{i(x-t)})$$

in $\mathbf{L}^p(\mathbf{T})$. For each finite p, (S_t) is a strongly continuous group of isometries. There is a well developed theory of vectorial integration that gives meaning to (3.12), and proves results extending those of ordinary integration theory such as the statement of the lemma. We shall not rely on the general theory, but use the abstract formulation to suggest useful inequalities that (like the lemma above) can be proved directly. Problem 3 below is such a result.

Problems

1. Show that convolution in $\mathbf{L}^1(\mathbf{T})$ is associative and commutative. Show that $a_n(f * g) = a_n(f) a_n(g)$ for f and g in $\mathbf{L}^1(\mathbf{T})$.

2. Show that $\mathbf{L}^1(\mathbf{T})$ has no identity for convolution.

3. Show that if g has a continuous derivative on $\mathbf{T}$, the same is true of $f * g$, where f is any summable function.

4. Show that $f*g$ is continuous if f and g belong to dual Lebesgue spaces on $\mathbf{T}$ or on $\mathbf{R}$. [Let (S_t) be the translation operators defined by (3.13) on $\mathbf{T}$, or analogously on $\mathbf{R}$. Then $(S_t f)$ is a continuous mapping of $\mathbf{T}$ or of $\mathbf{R}$ into $\mathbf{L}^p$ if p is finite, for any f in that space. Apply the Hölder inequality.]

5. Let (e_n) be an approximate identity on $\mathbf{T}$, and g an element of $\mathbf{L}^\infty(\mathbf{T})$. Show that e_n*g tends to g in the weak* topology of $\mathbf{L}^\infty(\mathbf{T})$. That is, for any summable f, $e_n*g*f(1)$ tends to $g*f(1)$.

6. Prove the Principle enunciated above.

7. Let (e_n) be an approximate identity on $\mathbf{T}$. Show that $a_k(e_n)$ tends to 1 as n tends to ∞, for each k.

8. An approximate identity on $\mathbf{R}$ is defined just as on $\mathbf{T}$. State and prove an analogue of Fejér's theorem on $\mathbf{R}$. [$\mathbf{C}(\mathbf{T})$ is replaced by $\mathbf{C}_0(\mathbf{R})$, the space of continuous functions that tend to 0 at $\pm\infty$.]

9. Assume that there is an approximate identity on $\mathbf{T}$ consisting of trigonometric polynomials. Prove that trigonometric polynomials are dense in each space $\mathbf{L}^p(\mathbf{T})$, $1 \le p < \infty$. [The existence of such an approximate identity will be proved in Section 5. This is the usual way to prove the Parseval relation.]

10. Prove this fact, used above: if $\int |fh| \, d\sigma \le k$ for every h in $\mathbf{L}^q(\mathbf{T})$ with norm 1, then f is in $\mathbf{L}^p(\mathbf{T})$ and $\|f\|_p \le k$. (p and q are conjugate exponents.)

4. Unicity theorem; Parseval relation; Fourier-Stieltjes coefficients

Theorem 2. *If f is in $\mathbf{L}^1(\mathbf{T})$ and $a_n(f) = 0$ for all n, then $f = 0$.*

First suppose that f has a continuous derivative. By the corollary of Theorem 1, f is the sum of its Fourier series everywhere. This sum is 0 because all the coefficients of f are 0. The theorem is proved for such functions.

It is easy to construct an approximate identity (e_n) on $\mathbf{T}$ consisting of functions with continuous derivatives (Problem 1 below). Let f be any function in $\mathbf{L}^1(\mathbf{T})$. According to Problem 3 of the last section, $e_n * f$ has continuous derivative. Its coefficients are the products of the coefficients of e_n and those of f (Problem 1 of the last section), so they are all 0. Thus for each n, $e_n * f = 0$ everywhere. The convolution converges to f in $\mathbf{L}^1(\mathbf{T})$ by Fejér's theorem, so f is the null function of $\mathbf{L}^1(\mathbf{T})$, as we wished to prove.

Since $\mathbf{L}^1(\mathbf{T})$ contains $\mathbf{L}^p(\mathbf{T})$ for $p > 1$, the same unicity statement holds in all the Lebesgue spaces.

Theorem 2 implies, for any finite p, that trigonometric polynomials are dense in $\mathbf{L}^p(\mathbf{T})$. For let q be the exponent conjugate to p. (If $p = 1$, then $q = \infty$.) If the assertion were false for $\mathbf{L}^p(\mathbf{T})$, there would be a nonnull element g of $\mathbf{L}^q(\mathbf{T})$ such that

$$(4.1) \qquad \int g(e^{ix}) h(e^{-ix}) \, d\sigma(x) = 0$$

for every trigonometric polynomial h. This means that the Fourier coefficients of g are all 0, so g is null, contrary to hypothesis.

In Section 1, the observation was made that Bessel's inequality is equality for every trigonometric polynomial, and in consequence for every function that is the limit of trigonometric polynomials in $\mathbf{L}^2(\mathbf{T})$. Now we have shown that every function in $\mathbf{L}^2(\mathbf{T})$ is such a limit, so the Parseval relation is now proved.

A famous theorem of Weierstrass asserts that the trigonometric polynomials are dense in $\mathbf{C}(\mathbf{T})$. This fact is not quite proved. If we knew that there is an approximate identity (e_n) consisting of trigonometric polynomials, then Fejér's theorem would give the result immediately (and the same is true for the spaces $\mathbf{L}^p(\mathbf{T})$ just treated). Such an approximate identity will be constructed in the next section, but first let us prove Weierstrass' theorem by an argument like the one used above.

A fundamental theorem of F. Riesz asserts that the dual of $\mathbf{C}(\mathbf{T})$ is $\mathbf{M}(\mathbf{T})$, the space of bounded complex Borel measures on $\mathbf{T}$ (realized as the interval $[0, 2\pi)$), where the value of μ as a functional on a continuous function h is

$$(4.2) \qquad \int h(e^{-ix})\, d\mu(x).$$

The *Fourier-Stieltjes coefficients* of μ are defined to be

$$(4.3) \qquad \hat{\mu}(n) = a_n(\mu) = \int e^{-nix}\, d\mu(x),$$

the value of μ as a functional on χ^n. This functional vanishes on all trigonometric polynomials if all its Fourier-Stieltjes coefficients are 0. If we show that then μ must be the null measure, it will follow that trigonometric polynomials are dense in $\mathbf{C}(\mathbf{T})$. Since $\mathbf{M}(\mathbf{T})$ contains $\mathbf{L}^1(\mathbf{T})$ (when we identity a function f with the measure $f\,d\sigma$), this is a stronger statement than Theorem 2.

Paradoxically, we prove this stronger statement from the unicity theorem for $\mathbf{C}(\mathbf{T})$, the weakest version of all. (But we are using the power of Riesz' theorem as well.) Let μ be a measure whose Fourier-Stieltjes coefficients are all 0. For h in $\mathbf{C}(\mathbf{T})$ define the convolution

$$(4.4) \qquad \mu * h(e^{ix}) = \int h(e^{i(x-t)}) \, d\mu(t).$$

This is a continuous function (because h is uniformly continuous). Note that (4.2) can be written $\mu * h(1)$. The Fourier coefficients of $\mu * h$ are

$$(4.5) \qquad a_n(\mu * h) = \int \int e^{-nix} h(e^{i(x-t)}) \, d\mu(t) \, d\sigma(x).$$

When we change the order of integration and replace x by $x + t$ we find

$$(4.6) \qquad a_n(\mu * h) = a_n(\mu) a_n(h).$$

By assumption the coefficients of μ vanish; hence $\mu * h$ has null coefficients, and is zero by the unicity theorem in $C(T)$. In particular $\mu * h(1) = 0$. Since h was an arbitrary function in $C(T)$, μ is the null functional, so by the Riesz theorem is the null measure, as we wanted to show.

The logical connections among all these results are complicated. We have to know, or to prove, that trigonometric polynomials are dense in $C(T)$ (the theorem of Weierstrass), and that $C(T)$ is dense in each space $L^p(T)$ $(1 \le p < \infty)$. The first statement is equivalent to the unicity theorem for measures, if we know the Riesz theorem; or it follows from the existence of an approximate identity consisting of trigonometric polynomials (which we will construct presently), using the first part of Fejér's theorem. However the other part of Fejér's theorem, about functions of the spaces $L^p(T)$, depends on the fact that $C(T)$ is dense in these spaces. A proof can be given using Lebesgue's theorem on the differentiation of indefinite integrals (a difficult result), or the fact may be obvious if $L^p(T)$ was defined in measure theory as the completion in some sense of $C(T)$. But here once again the Riesz theorem removes the difficulty, because the unicity theorem for $L^q(T)$ (where q is conjugate to p) shows that trigonometric polynomials are dense in $L^p(T)$.

The convolution of a measure with a continuous function has been defined, and incidentally it is shown that the Fourier coefficients of the convolution are the products of the coefficients of the factors. This definition enables us to define the convolution of two measures μ and ν as well. $\mu*\nu$ is to be that measure whose value as a linear functional on h in $C(T)$ is $[\mu*(\nu*h)](1)$. Obviously this double convolution determines a functional on $C(T)$ that is linear in the algebraic sense. Moreover

$$(4.7) \qquad |[\mu*(\nu*h)](1)| \leq ||\mu|| \, ||\nu*h||_\infty,$$

where the norm of μ means its norm as a linear functional on $C(T)$. From (4.4) it is obvious that $||\nu*h||_\infty \leq ||\nu|| \, ||h||_\infty$. Thus

$$(4.8) \qquad |[\mu*(\nu*h)](1)| \leq ||\mu|| \, ||\nu|| \, ||h||_\infty.$$

Hence this functional is continuous, with bound at most $||\mu|| \, ||\nu||$. This defines $\mu*\nu$, and shows that

$$(4.9) \qquad ||\mu*\nu|| \leq ||\mu|| \, ||\nu||.$$

The definition means that $[\mu*(\nu*h)](1) = [(\mu*\nu)*h](1)$ for all h in $C(T)$. The point 1 is not exceptional; by translating h we see that $(\mu*\nu)*h = \mu*(\nu*h)$ everywhere. (Both sides of the equality are continuous functions.)

From (4.4) we verify that $\mu*\chi^n = a_n(\mu)\chi^n$ for each n. Thus

$$a_n(\mu*\nu)\chi^n = (\mu*\nu)*\chi^n = \mu*(\nu*\chi^n) = a_n(\nu)\mu*\chi^n = a_n(\mu)\,a_n(\nu)\chi^n,$$

so that

$$(4.10) \qquad\qquad a_n(\mu*\nu) \;=\; a_n(\mu)\,a_n(\nu).$$

This fact, together with the unicity theorem, implies that convolution of measures is associative and commutative.

If one of the measures is absolutely continuous, it is possible to define the convolution directly by a formula like (4.4). However there are difficulties, and we have avoided them by the procedure used here.

$\mathbf{M(T)}$ is a commutative Banach algebra. It contains $\mathbf{L^1(T)}$, when a function f is identified with the measure $f d\sigma$. This algebra has an identity δ: unit mass carried at the point 1, with transform $\hat{\delta}(n) = 1$ for all n.

If f is continuous, then $\mu*f$ has been defined in two ways, by (4.4) and in a more complicated way when f is regarded as a measure. One can prove (Problem 2 below) that when f is a continuous function, $\mu*(f d\sigma)$ is an absolutely continuous measure $h d\sigma$, the summable function h is in fact continuous (that is, equal a.e. to a continuous function $\tilde{h}$), and $\tilde{h} = \mu*f$.

The last result of this section is a result of N. Wiener about Fourier-Stieltjes coefficients.

Wiener's theorem. *If the measure μ on $\mathbf{T}$ has point masses $a_1, a_2, \ldots,$ then*

$$(4.11) \qquad \lim \frac{1}{2N+1} \sum_{-N}^{N} |\hat{\mu}(n)|^2 = \sum |a_k|^2.$$

Define the measure ν by setting $\nu(E) = \bar{\mu}(-E)$ for each Borel set E of $\mathbf{T}$. Then $\hat{\nu}$ is the complex conjugate of $\hat{\mu}$. Hence

$$(4.12) \qquad \frac{1}{2N+1} \sum_{-N}^{N} |\hat{\mu}(n)|^2 = \int \frac{1}{2N+1} \sum_{-N}^{N} e^{-nix} \, d(\mu * \nu)(x).$$

As N tends to ∞, the integrand h_N tends to 1 at 0, to 0 elsewhere on $(-\pi, \pi)$, and its modulus never exceeds 1. Hence the limit on the right exists, and is the mass carried by $\mu * \nu$ at 0. We shall calculate this mass.

The right side of (4.12) equals $(\mu * \nu) * h_N(1)$, or

$$(4.13) \qquad \nu * (\mu * h_N)(1) = \int \int h_N(e^{i(y-x)}) \, d\mu(x) \, d\bar{\mu}(y).$$

Write $\mu = \mu_d + \mu_c$, the sum of the discrete and the continuous parts of μ. The inner integral with μ_c in place of μ tends to 0 for every y, because the integrand (a function of x) tends to 0 except at the point $x = y$ (a null set for μ_c). Hence

$$(4.14) \qquad \int \int h_N(e^{i(y-x)}) \, d\mu_c(x) \, d\bar{\mu}(y)$$

tends to 0 as N tends to ∞. The inner integral with μ_d also tends to 0 except at the countably many points (y_j) where μ carries the masses (a_j); at such a point y_j the inner integral tends to a_j. Hence the limit of the iterated integral is

$$(4.15) \qquad \int q(y) \, d\bar{\mu}(y),$$

where q is the function taking the value a_j at each point y_j and 0 at other points. The value of this integral is the right side of (4.11), and the theorem is proved.

If μ is a continuous measure, that is if it has no point masses, then the limit in (4.11) is 0. This statement is weaker than the conclusion of Mercer's theorem, that $\hat{\mu}(n)$ tends to 0 if μ is absolutely continuous.

Problems

1. Show that there is an approximate identity on **T** consisting of continuously differentiable functions.

2. Show that if μ is a measure and f a continuous function on **T**, $\mu*(f\,d\sigma) = h\,d\sigma$, where h is a.e. equal to the continuous function $\mu*f$. [Test $\mu*f$ as a linear functional on **C(T)**.]

3. Show that if μ is a measure and (e_n) an approximate identity on **T**, then $(\mu*e_n)$ tends to μ in the weak∗ topology of **M(T)** as the dual of **C(T)**.

4. Show that if the Fourier-Stieltjes sequence of a measure is in l^2, then the measure is absolutely continuous.

5. Show that the norm of f in $\mathbf{L^1(T)}$ is the same as its norm as a measure $f\,d\sigma$, that is, as a linear functional on **C(T)**.

6. Evaluate the constants

$$\int_0^{2\pi} \cos^{2n}x\,dx \quad (n = 1, 2, \ldots).$$

[$\exp 2\cos x = |\exp \chi|^2$. Write each exponential as a power series, then integrate.]

7. Sum the series $\sum_1^\infty 1/n^2$, $\sum_1^\infty 1/(2n-1)^2$, $\sum_1^\infty 1/(n^2+1)$ by applying the Parseval relation to appropriate functions. [One

function is $\exp(-x)$ on $(0, 2\pi)$.]

8. The identity $\dfrac{1}{\sin^2 x} = \sum\limits_{-\infty}^{\infty} \dfrac{1}{(x - n\pi)^2}$, valid for all complex x not real integer multiples of π, is usually proved by complex variable methods. Prove it (for real x) by applying the Parseval relation to the function $\exp iuy$ (y in $(-\pi, \pi)$) and then setting $u = x/\pi$.

5. The classical kernels

The *Fejér kernel* and the *Poisson kernel* are two particular approximate identities on **T** (or, with analogous definitions, on **R**) with agreeable and important properties. These kernels are tools in the theory of summability of Fourier series. The *Dirichlet kernel* plays the same rôle in studying the convergence of Fourier series. The fact that the Dirichlet kernel is not an approximate identity makes the theory of convergence more difficult and less satisfactory than that of summability.

The Dirichlet kernel (D_n) is defined by

$$(5.1) \qquad\qquad D_n(e^{ix}) = \sum_{-n}^{n} e^{kix}$$

for $n = 0, 1, \ldots$. This is a geometric series, and we find for its sum

$$(5.2) \qquad\qquad D_n(e^{ix}) = \frac{\sin(n + \frac{1}{2})x}{\sin \frac{1}{2}x} .$$

From (5.1) we see that

$$(5.3) \qquad\qquad D_n * f(e^{ix}) = \sum_{-n}^{n} a_k(f) e^{kix}.$$

(Convolution of functions multiplies their Fourier coefficients; therefore (5.3) is obvious as a Fourier relation. The fact that it is actually equality is checked by performing the integration that defines the convolution, using the expression (5.1) for D_n.) Since D_n is an even function, the Fourier series of f converges at 1 to the value

$$(5.4) \qquad \lim_{n \to \infty} \int D_n(e^{ix}) f(e^{ix}) \, d\sigma(x)$$

if this limit exists.

From (5.1) we have $a_0(D_n) = 1$ for each n. But (5.2) shows that the D_n are not of one sign, so they do not form an approximate identity. Indeed the numbers

$$(5.5) \qquad \int |D_n| \, d\sigma,$$

called the *Lebesgue constants*, tend to ∞ (Problem 1 below).

From the fact that the Lebesgue constants are unbounded we can prove that *there is a continuous function whose Fourier series diverges at the point* 1 (or at any other point), without actually constructing such a function. If it were true that

$$(5.6) \qquad S_n(h) = \sum_{-n}^{n} a_k(h) = D_n * h(1)$$

has a limit for every h in $\mathbf{C(T)}$, then (S_n) would be a sequence of linear functionals on $\mathbf{C(T)}$ that is bounded at each element of the space. The Banach-Steinhaus theorem asserts in this case that the sequence of norms $(\|S_n\|)$ is bounded. But the norm of this functional is the same as $\|D_n\|_1$, which tends to ∞. The contradiction establishes the result. (We have proved more: $S_n(h)$ must be unbounded for some h.)

For many years it was not known whether the Fourier series of a continuous function must converge *somewhere*. Kolmogorov gave an example of a function in $L^1(T)$ whose Fourier series diverges everywhere. Only much later, in 1966, L. Carleson proved that the Fourier series of every function in $L^2(T)$ converges almost everywhere, and this is true in particular for continuous functions. R. Hunt showed that the same is true in $L^p(T)$ for $p > 1$.

The Fejér kernel is

$$(5.7) \qquad K_n(e^{ix}) = \sum_{-n}^{n} \left(1 - \frac{|k|}{n}\right) e^{kix}$$

for $n = 1, 2, \ldots$. Thus $a_0(K_n) = 1$, and the coefficients decrease linearly to 0 at $\pm n$. This series can be summed directly, but an indirect procedure is easier. If we multiply out the square in

$$(5.8) \qquad \frac{1}{n} \left| \sum_{0}^{n-1} e^{kix} \right|^2$$

we find exactly the coefficients of K_n, and so this is K_n. When we sum the geometric series and simplify, we find

$$(5.9) \qquad K_n(e^{ix}) = \frac{1}{n} \left(\frac{\sin \frac{1}{2} nx}{\sin \frac{1}{2} x} \right)^2.$$

Thus the Dirichlet and Fejér kernels are related by the formula

$$(5.10) \qquad K_{2n+1} = \frac{1}{2n+1} D_n^2.$$

From (5.9) we see that K_n is positive everywhere, and tends to 0 uniformly on $(-\pi, \pi)$ outside any interval about 0. Furthermore K_n has mean value 1, so (K_n) is an approximate identity.

The functions K_n are trigonometric polynomials. We have already mentioned some of the consequences of the fact that there is such an approximate identity. A more practical result is this: we want to approximate a function f in some Lebesgue space by trigonometric polynomials. A theorem about the density of trigonometric polynomials tells us the approximation is possible; but the Fejér kernel provides an effective way of writing down such trigonometric polynomials, namely $K_n * f$.

The Poisson kernel is a family (P_r) depending on the continuous parameter r, which varies from 0 to 1. (The definition of an approximate identity should be phrased in such generality as to apply to a general parameter set, directed towards a limit point. This would be so complicated that we prefer to present the idea and not a general definition. In this case r increases to 1.) We define

$$(5.11) \qquad P_r(e^{ix}) = \sum_{-\infty}^{\infty} r^{|n|} e^{nix} \qquad (0 < r < 1).$$

The series converges absolutely, and we compute that

$$(5.12) \qquad P_r(e^{ix}) = \frac{1 - r^2}{(1 - 2r\cos x + r^2)}.$$

From (5.11) we have $a_0(P_r) = 1$ for all r; (5.12) shows that P_r is non-negative, and tends to 0 uniformly as r increases to 1 outside an arbitrary interval about 1 on $\mathbf{T}$. Thus (P_r) is an approximate identity.

The Poisson kernel provides a harmonic extension of functions f in $\mathbf{L}^1(\mathbf{T})$ to the interior of the disk. If f has coefficients (a_n), then

$$(5.13) \qquad P_r*f(e^{ix}) = \sum_{-\infty}^{\infty} a_n \, r^{|n|} \, e^{nix}.$$

Each term satisfies Laplace's equation: replacing the angle x temporarily by θ, this means that

$$(5.14) \qquad \left(\frac{\partial^2}{\partial x^2} + \frac{\partial^2}{\partial y^2}\right) r^{|n|} \, e^{ni\theta} = 0.$$

Since (a_n) is a bounded sequence, the series in (5.13) converges uniformly on each disk interior to the unit circle. From function theory we know that the sum is harmonic in the open unit disk. Conversely, every function harmonic in the disk has an expansion like the right side of (5.13), although the coefficients need not be Fourier coefficients of a function defined on the boundary. (A complex function satisfies Laplace's equation if and only if its real and imaginary parts do. In this subject it is natural to allow harmonic functions to take complex values, although problems about them are usually reduced to the case of real functions.)

Since (P_r) is an approximate identity, P_r*f converges to f in the metric of $\mathbf{L}^p(\mathbf{T})$ if f belongs to that space (p finite). This is the same as saying that the Fourier series of f is summable to f by Abel means. This is the boundary-value theorem for metric convergence. It is also true that P_r*f converges to f almost everywhere, but this result is more difficult to prove, it depends on particular properties of the kernel, and is not true for all approximate identities.

If $a_n(f) = 0$ for all $n < 0$, then (5.13) is a power series in $z = re^{ix}$ and $F(z) = P_r*f(e^{ix})$ is analytic in the open unit disk. The subspace of $\mathbf{L}^p(\mathbf{T})$ consisting of all such f is called $\mathbf{H}^p(\mathbf{T})$. We shall study these spaces in the sequel.

Problems

1. Show that the Lebesgue constants tend to ∞. Estimate them above and below as closely as possible.

2. Carry through the proof of (5.14).

3. Prove the corollary to Theorem 1 anew by means of (5.2) and (5.3).

4. Verify that for r and s between 0 and 1, $P_{rs} = P_r*P_s$. Deduce that for f in $\mathbf{L}^p(\mathbf{T})$, $||P_r*f||_p$ is an increasing function of r $(0 < r < 1)$.

5. Suppose that f is summable and non-negative. Show that f is the boundary function, in the sense of this section, of a non-negative harmonic function. If f is bounded, the harmonic function has the same bounds.

6. Show that if f is real and bounded, and P_r*f assumes a value at an interior point of the disk that is a maximum (or a minimum) of its values in the whole open disk, then f is constant.

6. Summability: metric theorems

In the last section we saw that each function in $\mathbf{L}^1(\mathbf{T})$ has a harmonic extension to the interior of the unit disk given by

(6.1) $$F(re^{ix}) = P_r*f(e^{ix}).$$

More generally, any finite complex measure μ has a similar extension obtained by convoluting with P_r. Let f_r be the function on $\mathbf{T}$
defined by $f_r(e^{ix}) = F(re^{ix})$. Each f_r is continuous, but we choose
to consider it as an element of $\mathbf{L}^1(\mathbf{T})$. The family (f_r) is bounded
in $\mathbf{L}^1(\mathbf{T})$, because by (4.9)

$$(6.2) \qquad ||P_r * \mu||_1 \leq ||P_r||_1 \, ||\mu|| = ||\mu||.$$

Conversely, given such a family (f_r) bounded in $\mathbf{L}^1(\mathbf{T})$, we
would like to find a boundary function or measure that generates
the family by convolution with the Poisson kernel.

Theorem 3. *Let F be a function harmonic in the unit disk
that satisfies*

$$(6.3) \qquad A_r = \int |F(re^{ix})| \, d\sigma(x) \leq K$$

*for $0 < r < 1$. Then F is the Poisson integral of a uniquely
determined measure μ, and $||\mu|| = \lim\limits_{r \uparrow 1} A_r$.*

The harmonic function F is the sum of a series

$$(6.4) \qquad F(re^{ix}) = \sum a_n r^{|n|} e^{nix}$$

convergent in the disk. The functions f_r are defined from F as
above, and obviously $a_n(f_r) = a_n r^{|n|}$. This family is bounded in
$\mathbf{L}^1(\mathbf{T})$, and therefore bounded as elements of the dual of $\mathbf{C}(\mathbf{T})$. A
closed ball in the dual of a Banach space is compact for the weak$*$
topology (this is the theorem of Banach and Alaoglu). Hence
there is a measure μ with the property that every $*$-neighborhood
of μ contains $f_r d\sigma$ for values of r arbitrarily close to 1. Given an

exponential e^{nix} and a positive number ϵ, the measures ν such that

$$(6.5) \qquad \left| \int e^{-nix} \, d\nu(x) - \int e^{-nix} \, d\mu(x) \right| < \epsilon$$

constitute a $*$-neighborhood of μ. Therefore $a_n(f_r)$ is as close as we please to $a_n(\mu)$ for a sequence of values of r tending to 1. Since $a_n(f_r)$ tends to a_n as r increases to 1, we must have $a_n(\mu) = a_n$. This is true for all integers n. Therefore $f_r = P_r * \mu$, which is the first statement of the theorem.

By Problem 3 of Section 4, μ is the $*$-limit of $f_r d\sigma$. Also A_r increases with r (Problem 4 of the last section). It follows that $\|\mu\| \le \lim_{r \uparrow 1} A_r$. If the inequality were strict, we should have $\|f_s\|_1 \le \|\mu\| < \lim_{r \uparrow 1} \|f_r\|$ for each s, $0 < s < 1$, which is impossible.

The Fourier-Stieltjes coefficients of μ, and therefore μ itself, are uniquely determined by F, because $a_n(\mu) = \lim_{r \uparrow 1} a_n(f_r)$.

Corollary. *Suppose that F is harmonic in the unit disk and satisfies*

$$(6.6) \qquad A_r^p = \int |F(re^{ix})|^p \, d\sigma(x) \le K < \infty, \qquad 0 < r < 1$$

for some $p > 1$. Then F is the Poisson integral of a unique function f in $\mathbf{L}^p(\mathbf{T})$, and $\|f\|_p = \lim_{r \uparrow 1} A_r$. For $p = \infty$ the same conclusion holds if F is bounded and A_r is the bound of f_r.

The proof is like that of the theorem except for a simplification: now $\mathbf{L}^p(\mathbf{T})$ is itself a dual space, so (f_r) has a weak$*$ limit in $\mathbf{L}^p(\mathbf{T})$, and F is the Poisson integral of this function.

If $1 < p < \infty$, f_r converges to f not only in the weak$*$

topology (which is the same as the weak topology for these values of p, because the spaces are reflexive), but in norm. For once we know that $f_r = P_r * f$, Fejér's theorem asserts norm convergence. The same is true for $p = 1$ if the limiting measure μ is absolutely continuous, but of course not otherwise. For $p = \infty$ we have norm convergence if and only if f is continuous.

The theorem establishes a correspondence between Fourier series, which represent functions or measures on the boundary **T** of the disk, and harmonic functions on the disk. This correspondence will be strengthened by theorems asserting pointwise convergence or summability in the following sections. In the classical function theory of the disk various results that can be phrased and proved in terms of Fourier series are treated instead, by means of these boundary value theorems, as results about analytic or harmonic functions. This confusion hides the added interest of those results that really do depend on function theory.

The limit theorems just proved have not used any property of the kernel except that it is an approximate identity. Let (e_n) be any approximate identity, with Fourier coefficients λ_{nk}. Suppose for each n the series

$$(6.7) \qquad \sum_{k=-\infty}^{\infty} a_k \lambda_{nk} e^{kix}$$

converges absolutely to a function f_n, and suppose the sequence (f_n) is bounded in $\mathbf{L}^p(\mathbf{T})$ for some p. Then there is a function f in $\mathbf{L}^p(\mathbf{T})$ whose Fourier coefficients are (a_k), unless $p = 1$; in that case the limiting object may be a measure.

When the approximate identity is the Fejér kernel, the sums (6.7) are called *Cesàro means* of the series

(6.8)
$$\sum a_k e^{kix},$$

and if f_n converges to f in $\mathbf{L}^P(\mathbf{T})$, (6.8) is said to be *summable by Cesàro means to f* in that space. The process associated with the Poisson kernel is called *Abel summability.*

This metric summability is simple and general. Pointwise summability, that is, pointwise convergence of f_n, depends on special properties of the kernels. This is the subject of the next section.

Problems

1. Let F be a non-negative harmonic function in the unit disk. Show that (6.6) holds with $p = 1$, so that F is the Poisson integral of a positive measure on $\mathbf{T}$.

2. Say we know that every function analytic in the unit disk has a Taylor series convergent in the disk. Explain why every complex harmonic function in the disk has a development (6.4).

3. Show that a measure μ is absolutely continuous if and only if $(P_r * \mu)$ is a Cauchy sequence in $\mathbf{L}^1(\mathbf{T})$. Deduce that $\mathbf{L}^1(\mathbf{T})$ is an ideal in $\mathbf{M}(\mathbf{T})$.

4. In general topology, a sequence (x_n) may have an accumulation point x, and yet x may not be the limit of any subsequence of (x_n). This possibility complicated a proof given in this section. However, there really is no difficulty, because of this result: *If B is a separable Banach space, (x_n) a bounded sequence in the dual of B, and x an accumulation point of the sequence for the weak* topology, then a subsequence of (x_n) converges weak* to* x. Prove this result, and use it to simplify the proof above. [Recall the Principle of Section 3. To say that x is an accumulation point

of the sequence means that every neighborhood of x contains x_n for infinitely many n.]

7. Pointwise summability

For any approximate identity (e_n) on $\mathbf{T}$ and f in $\mathbf{L}^1(\mathbf{T})$, $e_n * f$ converges to f in norm, and therefore a subsequence converges to f almost everywhere. Unfortunately the subsequence depends on f. It is an important fact that the Poisson and Fejér kernels give convergence a.e. without passing to a subsequence. This depends on particular properties of these kernels, and is harder to prove than the theorem on metric convergence.

Theorem 4. *For f in $\mathbf{L}^1(\mathbf{T})$, $P_r * f$ converges to f almost everywhere as r increases to 1. More exactly, the convolution converges to*

$$(7.1) \qquad \lim_{t \downarrow 0} (2t)^{-1} \int_{-t}^{t} f(e^{i(x-u)})\, du$$

at each point where this limit exists.

If $F(x)$ is an indefinite integral of $f(e^{ix})$, then (7.1) exists and equals $F'(x)$ at each point where this derivative exists. By a classical theorem of Lebesgue, the derivative exists and equals $f(e^{ix})$ almost everywhere.

On reflection the theorem is not very surprising. Convolution with an approximate identity is an averaging process, replacing $f(e^{ix})$ by an average of values of f near e^{ix}. The expression (7.1) is the convolution of f with the approximate identity taking the value $1/2t$ on $(-t, t)$, and 0 elsewhere on $(-\pi, \pi)$. The theorem asserts that one averaging process has a limit at each point where another one does. This kind of implication is called a *Tauberian theorem.*

We shall prove the theorem at $x = 0$, assuming that the limit (7.1) exists there. Adding a constant to f merely adds the same constant to the value of each convolution; thus we may assume that the limit in (7.1) is 0, and we are to show that $P_r * f$ converges to 0. Finally, we may suppose that f has mean value 0. For let g be a smooth function with the same mean value as f, and vanishing on a neighborhood of 1. Then (7.1), with g for f, equals 0, and also $P_r * g$ tends to 0. If we prove the theorem for $f - g$, then it follows for f.

Lemma. $r^{-1}(-\sin x P_r'(e^{ix}))$ *is an approximate identity on* **T**. *(The prime denotes differentiation with respect to x.)*

$P_r(e^{ix})$ is an even function, so its derivative is odd. We check from (5.12) that it decreases on $(0, \pi)$. Therefore the expression of the lemma is non-negative everywhere. Its mean value can be found by differentiating (5.11), multiplying by $(e^{ix} - e^{-ix})/2i$, and identifying the central coefficient. Finally we must show that for $0 < \epsilon < \pi$ we have

$$(7.2) \qquad \lim_{r \uparrow 1} \int_{-\epsilon}^{\epsilon} \sin x \, P_r'(e^{ix}) \, dx = -2\pi.$$

Integration by parts gives for the integral

$$(7.3) \qquad 2 \sin \epsilon \, P_r(e^{i\epsilon}) - \int_{-\epsilon}^{\epsilon} \cos x \, P_r(e^{ix}) \, dx.$$

The first term tends to 0 as r increases to 1 by (5.12), and the limit of the second is 2π from the definition of an approximate identity. This proves the lemma.

Define

$$(7.4) \qquad F(t) = \int_{-\pi}^{t} f(e^{ix})\, dx.$$

Since f has mean value 0, $F(-\pi) = F(\pi) = 0$. We have

$$(7.5) \quad 2\pi P_r * f(1) = \int_{-\pi}^{\pi} P_r(e^{-ix}) f(e^{ix})\, dx = \int_{-\pi}^{\pi} P_r'(e^{-ix}) F(x)\, dx,$$

where the product term in the integration by parts vanished. By the fact that P_r' is odd, the right side can be written

$$(7.6) \qquad \int_{0}^{\pi} 2x P_r'(e^{-ix}) \frac{F(x) - F(-x)}{2x}\, dx.$$

We recognize the fraction as the expression in (7.1) whose limit at 0 is 0. Since F is bounded, the fraction is bounded over the interval. The lemma shows that (7.6) has limit 0 as r incre ,ses to 1, and the theorem is proved.

Theorem 5. *For f in $\mathbf{L}^1(\mathbf{T})$, $K_n * f(e^{ix})$ tends to L at each point e^{ix} such that*

$$(7.7) \qquad \lim_{t \downarrow 0} t^{-1} \int_{0}^{t} |f(e^{i(x+u)}) + f(e^{i(x-u)}) - 2L|\, du = 0.$$

For almost every x this is the case with $L = f(e^{ix})$.

This theorem is not very different from the previous one, but the essential property of the kernel is not so natural to formulate and the proof is not so neat. We shall use these facts:

$$(7.8) \qquad K_n(e^{ix}) \leq n \quad \text{and} \quad \leq \frac{\pi^2}{nx^2}$$

for all n, and all x in $(0, \pi)$. These inequalities are easily verified from (5.8, 5.9).

Set $q(e^{ix}) = f(e^{i(x+u)}) + f(e^{i(x-u)}) - 2L$. We have

(7.9) $$2\pi(K_n * f(e^{ix}) - L) = \int_{-\pi}^{\pi} K_n(e^{it})\,(f(e^{i(x-t)}) - L)\,dt$$
$$= \int_{0}^{\pi} K_n(e^{it})\,q(e^{it})\,dt$$

because K_n is even. For positive t define

(7.10) $$H(t) = \int_{0}^{t} |q(e^{iu})|\,du.$$

By hypothesis, $H(t) = o(t)$ as t tends to 0.

The right side of (7.9) has modulus at most

(7.11) $$\int_{0}^{2\pi/n} K_n(e^{it})\,|q(e^{it})|\,dt + \int_{2\pi/n}^{\pi} K_n(e^{it})\,|q(e^{it})|\,dt = I + J.$$

Since $K_n \le n$,

(7.12) $$I \le nH(2\pi/n),$$

which tends to 0.

To estimate J we use the other bound from (7.8), and then integrate by parts:

(7.13) $$J \le \pi^2 n^{-1} \int_{2\pi/n}^{\pi} t^{-2}\,|q(e^{it})|\,dt =$$
$$\pi^2 n^{-1}\Big\{[t^{-2}H(t)]_{2\pi/n}^{\pi} + 2\int_{2\pi/n}^{\pi} t^{-3}H(t)\,dt\Big\}.$$

The first term in braces, divided by n, tends to 0 by the hypothesis on H. If we replace t by t/n, the second term in braces with its factor becomes

(7.14)
$$2\pi^2 n \int_{2\pi}^{n\pi} t^{-3} H(t/n) \, dt,$$

which tends to 0 by the dominated convergence theorem (Problem 5 below). This completes the proof of the first part of the theorem.

The fact that the condition holds with $L = f(e^{ix})$ at almost every point follows immediately from this theorem of Lebesgue: *for f in* $\mathbf{L}^1(\mathbf{T})$

(7.15)
$$\int_0^t |f(e^{i(x-u)}) - f(e^{ix})| \, du = o(t)$$

for almost all x. This is by no means obvious, but it is not our subject.

Note that (7.7) is a stronger condition than the existence of (7.1) (Problem 1 below).

The theorems on radial limits of harmonic functions were published in an extraordinary paper by P. Fatou in 1906.

Problems

1. Show that if (7.7) holds, then the limit (7.1) exists and equals L.

2. It would be convenient to prove Theorem 5 by showing that $-\sin(x/2)K_n'(e^{ix})$ has properties like those of an approximate identity. Show that, on the contrary, this family is essentially the Dirichlet kernel. [The prime means differentiation with respect to x. Note that $\sin(x/2)K_n'(e^{ix}) = (\sin(x/2)K_n(e^{ix}))' - \frac{1}{2}\cos(x/2)K_n.$]

3. If f is in $\mathbf{L}^1(\mathbf{T})$ and vanishes on an interval J of $\mathbf{T}$, then P_r*f and K_n*f converge to 0 uniformly on each interval interior to J.

4. Adapt the proof of Theorem 4 to show: if μ is a singular measure on $\mathbf{T}$, then $P_r*\mu$ converges to 0 a.e. as r increases to 1.

5. The proof that (7.14) tends to 0 is not quite trivial. Explain how the dominated convergence theorem applies.

8. Positive definite sequences; Herglotz' theorem

A complex sequence (u_n) is called *positive definite* if

$$(8.1) \qquad \sum_{m,n} u_{m-n} c_m \bar{c}_n \geq 0$$

for every sequence (c_n) such that $c_n = 0$ except for finitely many n. (Thus the sum in (8.1) is a finite one, and the property of being positive definite is purely algebraic.)

There is a way to generate positive definite sequences. Let μ be any *positive* measure on $[0, 2\pi)$. Set

$$(8.2) \qquad u_n = \int e^{-inx}\, d\mu(x).$$

Then

$$(8.3) \qquad \sum u_{m-n} c_m \bar{c}_n = \int |\sum c_n e^{-inx}|^2\, d\mu(x) \geq 0.$$

The result to be proved is that these are the only positive definite sequences.

Theorem of Herglotz. *Every positive definite sequence (u_n) is the Fourier-Stieltjes sequence of a positive measure.*

This beautiful theorem has generalizations that are important throughout modern analysis.

Choose $c_n = e^{nit}$ for $n = 0, 1, \ldots, N-1$, where N is a positive integer and t a real number, and $= 0$ for other n. A simple calculation shows that

$$(8.4) \qquad \sum u_{m-n} c_m \bar{c}_n = N \sum_{-N}^{N} u_n e^{nit} \left(1 - \frac{|n|}{N}\right),$$

and this expression is non-negative for every t. Let P_N be the sum in (8.4). The norm of P_N in $\mathbf{L}^1(\mathbf{T})$ is its mean value, equal to u_0 for each N. Since the family (P_N) is bounded, it has an accumulation point in the dual of $\mathbf{C}(\mathbf{T})$, which is a measure μ by the theorem of F. Riesz. The argument of Section 6 shows that μ has Fourier-Stieltjes sequence (u_n). Moreover μ is positive as a functional and therefore is a positive measure. This completes the proof.

Problems

1. Show from the definition that a positive definite sequence (u_n) has these properties: $|u_n| \le u_0$, $u_{-n} = \bar{u}_n$ for all n.

2. Show that for any sequence (v_n) in $\mathbf{l}^2$ the sequence

$$u_n = \sum_m v_{m+n} \bar{v}_m$$

is positive definite. Identify the measure of (8.2).

3. Let (u_n) be positive definite. Show that the set of n such that $|u_n| = u_0$ is a subgroup of the integers.

4. Find conditions on numbers u_0, u_1, u_{-1} for them to be part of a positive definite sequence.

5. (a) Show that the pointwise limit of positive definite sequences is positive definite. (b) For each positive integer k, let (u_n^k) be the Fourier-Stieltjes sequence of a positive measure μ_k. Suppose u_n^k converges to u_n, for each n. Show that μ_k converges in

the ∗-topology of measures to a measure μ whose Fourier-Stieltjes sequence is (u_n).

9. The inequality of Hausdorff and Young

The relations

$$(9.1) \qquad \|\hat{f}\|_\infty \leq \|f\|_1, \qquad \|\hat{f}\|_2 = \|f\|_2$$

hold for any function f in $\mathbf{L}^1(\mathbf{T})$ or $\mathbf{L}^2(\mathbf{T})$, respectively. The first is trivial, and the second is the Parseval relation. (The norms on the left refer to spaces of sequences.) These are the extreme cases of the *inequality of Hausdorff and Young*:

$$(9.2) \qquad \|\hat{f}\|_q \leq \|f\|_p$$

for any f in $\mathbf{L}^p(\mathbf{T})$, where $1 \leq p \leq 2$, and q is the exponent conjugate to p.

No similar inequality is true if $p > 2$.

The inequality (properly enunciated) holds on any locally compact abelian group and its dual. In particular, a function whose Fourier coefficients are a sequence in $\mathbf{l}^p$ $(1 < p < 2)$ belongs to $\mathbf{L}^q(\mathbf{T})$. This is not a case of (9.2), but can be proved from it by duality (Problem 1 below).

The truth of (9.2) for $p = 1$ and 2 suggests its truth for intermediate values, and indeed the proof depends on a convexity argument that makes the result a consequence of the easy special cases. The convexity argument is an *interpolation theorem*, proved by M. Riesz and extended by G. O. Thorin to a powerful method. Analytic function theory is used to prove the theorem of

Riesz and Thorin, and the proof is not simple. We shall give a direct proof of the inequality due to A. P. Calderón and A. Zygmund, which uses the function-theoretic ideas that enter into the proof of the theorem of Riesz and Thorin.

It will suffice to prove this statement. *For each trigonometric polynomial f with Fourier coefficients (c_n) and norm equal to 1 in $\mathbf{L}^p(\mathbf{T})$ we have*

$$(9.3) \qquad \|c\|_q \le 1.$$

For this inequality to hold for all f, it is necessary and sufficient that

$$(9.4) \qquad |\sum c_n d_n| \le 1$$

for every sequence d with norm 1 in $\mathbf{l}^p$.

Write $f = F^{1/p} E$, $d_n = D_n^{1/p} e_n$, where F and D_n are non-negative, and E, e_n have modulus 1. Thus

$$(9.5) \qquad \int F\, d\sigma = 1, \qquad \sum D_n = 1.$$

The left side of (9.4) can be written

$$(9.6) \qquad \left|\sum D_n^{1/p} e_n \int F^{1/p} E \chi^{-n}\, d\sigma\right|.$$

Now replace $1/p$ by the complex variable z, to define the analytic function

$$(9.7) \qquad Q(z) = \sum D_n^z e_n \int F^z E \chi^{-n}\, d\sigma.$$

The sum has only finitely many terms, each one obviously bounded in the strip $1/2 \le \Re z \le 1$. Hence Q is bounded in this strip. We seek a bound better than the obvious one obtained by estimating each term of the sum.

For $\Re z = 1$ we have the trivial estimate

$$(9.8) \qquad |Q(1 + it)| \le \sum D_n \int F \, d\sigma = 1.$$

For $\Re z = 1/2$ the Schwarz inequality gives

$$(9.9) \quad |Q(\tfrac{1}{2} + it)| \le \left(\sum D_n\right)^{1/2} \left(\sum \left|\int F^{(1/2)+it} E \chi^{-n} \, d\sigma\right|^2\right)^{1/2}.$$

The integral is the Fourier coefficient of $F^{(1/2)+it} E$. Bessel's inequality shows that the last square root is at most equal to $\|F^{1/2}\|_2 = 1$. Therefore the right side of (9.9) is at most 1.

The analytic function Q is bounded in the strip $1/2 \le \Re z \le 1$, and bounded by 1 on the lines $\Re z = 1/2, 1$. The maximum principle (Problem 2 below) asserts that $|Q(z)| \le 1$ throughout the strip. Taking $z = 1/p$ in (9.7) shows that the quantity (9.6) is at most equal to 1, as we wanted to show.

Problems

1. Show that if f is a summable function whose coefficient sequence is in $\mathbf{l}^p$, $1 < p < 2$, then f is in $\mathbf{L}^q(\mathbf{T})$ and $\|f\|_q \le \|\hat{f}\|_p$. [We have shown that the mapping from f to $\hat{f}$ carries $\mathbf{L}^p(\mathbf{T})$ into $\mathbf{l}^q$ and reduces norm. Find the adjoint of this mapping.]

2. Prove the maximum modulus principle as needed in the proof above. [Map the strip conformally onto the unit disk, and use facts we know about the Poisson representation of bounded

analytic functions.]

3. Show that the inequality (9.2) holds in general if it holds for trigonometric polynomials.

4. Show that if f is a trigonometric polynomial and (9.1) is equality, with $1 < p < 2$, then f is a monomial. [The fact is true for all f, but the proof is more complicated.]

10. Measures with bounded powers; endomorphisms of l^1

$M(R)$ is the algebra of complex bounded Borel measures on the line. The measure with unit mass at 0 is an identity for this algebra, so it makes sense to speak of inverses. Let μ be the measure carrying mass ϵ at the point u, where ϵ is a number of modulus 1. Then μ has an inverse, the measure with mass $\bar{\epsilon}$ at $-u$. The power μ^n (in the sense of convolution) is the point measure with mass ϵ^n at nu, for all integers n; these measures all have total mass 1. A theorem of A. Beurling and the author states that if μ is any measure such that $\|\mu^n\|$ is bounded over all integers n, then μ has the form just described: a unit mass on a single point of R.

The proof was complicated and depended on an arithmetic argument. J.-P. Kahane has found a simple and beautiful proof for the analogous theorem on the integer group, which we give now. (He has recently extended the proof to give the original theorem on the line, and also further results.)

The algebra is l^1, and a measure μ is a sequence such that $\sum |\mu(n)| < \infty$.

Theorem 6. *The element μ of l^1 has bounded convolution powers if and only if it satisfies $|\mu(n)| = 1$ for some n, $\mu(m) = 0$ for all $m \neq n$.*

Let μ have powers with norms bounded by a number K. The Fourier transform of μ is defined by

$$(10.1) \qquad m(e^{ix}) = \sum \mu(j)\, e^{-jix}.$$

Then the powers of m (with ordinary multiplication) are the transforms of the powers of μ (in the sense of convolution). Therefore

$$(10.2) \qquad |m(e^{ix})^n| \leq K \qquad \text{(all } n \text{ and } x);$$

hence $|m| = 1$ everywhere.

Since m is continuous, we can write $m(e^{ix}) = \exp i\phi(x)$ where ϕ is continuous on the line, and $\phi(x+2\pi) - \phi(x)$ is a constant multiple of 2π. The conclusion of the theorem is equivalent to the statement that $\phi(x) = px + b$ for an integer p and some real number b. Without loss of generality we suppose that $\phi(0) = 0$, and we shall prove that $\phi(x) = px$.

Lemma. *Given a positive number K, there is a positive number k such that any sequence μ with $||\mu||_1 \leq K$ and $||\mu||_2 = 1$ must satisfy $||\mu||_4 \geq k$.*

The proof (Problem 1 below) uses only the Hölder inequality.

The function

$$(10.3) \qquad \int e^{ni[\phi(t-s)+\phi(s)]}\, d\sigma(s)$$

is the convolution of $m^n = \exp ni\phi$ with itself, and this is the transform of $(\mu^n)^2$, where the nth power refers to convolution, and the square to ordinary multiplication. By Lemma 1 and the Parseval relation,

$$(10.4) \qquad 0 < k^4 \leq \sum_j |\mu^n(j)|^4 = \int \exp ni\Phi \, d\sigma(r, s, t),$$

where

$$(10.5) \qquad \Phi(r, s, t) = \phi(t-r) + \phi(r) - \phi(t-s) - \phi(s).$$

The integrand is periodic, and the measure is the product measure $d\sigma(r) \, d\sigma(s) \, d\sigma(t)$.

We shall show that $\exp i\Phi$ takes some constant value on a set of (r, s, t) of positive measure.

Let ν be the distribution measure of $\exp i\Phi$. That is, for E a Borel subset of $\mathbf{T}$, $\nu(E)$ is the normalized measure of the set of (r, s, t), $0 \leq r, s, t < 2\pi$, such that $\exp i\Phi(r, s, t)$ belongs to E. The basic definitions of integration theory show that the right side of (10.4) equals

$$(10.6) \qquad \int e^{niu} \, d\nu(u).$$

If $\exp i\Phi$ assumes each value only on a null set, then ν has no point masses. By Wiener's theorem of Chapter 1, Section 4,

$$(10.7) \qquad \lim N^{-1} \sum_{-N}^{N} |\hat{\nu}(n)|^2 = 0.$$

Thus $|\hat{\nu}(-n)|$ cannot exceed k^4 for all n, as asserted by (10.4). This contradiction shows that $\exp i\Phi$ assumes some value w on a set A of positive measure in $\mathbf{T}^3$.

The function

$$(10.8) \qquad \Psi = (1 + w^{-1} e^{i\Phi})/2$$

equals 1 on A, and has modulus less than 1 elsewhere on $\mathbf{T}^3$. Thus Ψ^n is 1 on A, and tends to 0 everywhere else. We want to show that Ψ^n has absolutely convergent Fourier series on $\mathbf{T}^3$, with coefficient norms that are bounded in n.

The function $\exp i\Phi$ has absolutely convergent Fourier series on $\mathbf{T}^3$, for

$$(10.9) \qquad e^{i\Phi} = \left(\sum \mu(j) \, e^{-ji(t-r)} \right) \left(\sum \mu(j) \, e^{-jir} \right)$$
$$\times \left(\sum \mu(j) \, e^{ji(t-s)} \right) \left(\sum \mu(j) \, e^{jis} \right).$$

Each factor is an absolutely convergent sum, and can be viewed as a Fourier series on $\mathbf{T}^3$. Since the $\mu(j)$ in modulus have sum at most K, the product in (10.9) has coefficients whose moduli have sum at most K^4. The same argument applies to powers of (10.9), and since the convolution powers of μ have norm at most K, the powers of (10.9) are absolutely convergent Fourier series with coefficient norms bounded by K^4.

Now take the nth power of (10.8) and expand by the binomial theorem. The coefficient bound for $\exp ni\Phi$ leads to the fact that Ψ^n also has coefficient norm at most K^4. The element of $\mathbf{l}^1(\mathbf{Z}^3)$ whose transform is Ψ will be called ψ. We have shown that ψ^n has norm at most K^4 in $\mathbf{l}^1(\mathbf{Z}^3)$ for each n.

Now $\mathbf{l}^1(\mathbf{Z}^3)$ is the dual space of $\mathbf{c}_0(\mathbf{Z}^3)$, the space of sequences (depending on three indices) that tend to 0 at ∞. (This means: arbitrarily close to 0 except for finitely many indices.)

Since the family (ψ^n) is bounded in $l^1(\mathbf{Z}^3)$, it has an accumulation point ρ for the weak∗ topology of the space. By the definition of this topology it follows that Ψ^n, at least on a subsequence, converges pointwise to the transform of ρ, which of course is a continuous function on $\mathbf{T}^3$. But the pointwise limit of Ψ^n is 1 on a set of positive measure, and 0 elsewhere. To be continuous, this limit must be 1 everywhere. That is, $\exp i\Phi = w$ everywhere.

The rest of the proof is straightforward. For some real number c,

$$(10.10) \qquad \phi(t{-}r) + \phi(r) - \phi(t{-}s) - \dot\phi(s) = c \quad (\text{mod } 2\pi)$$

for all r, s, t. Since ϕ is continuous, the equation holds literally everywhere. Taking all the variables equal to 0 shows that $c = 0$. Hence $\phi(t{-}r) + \phi(r)$ is independent of r, and (since $\phi(0) = 0$) equals $\phi(t)$. In other words

$$(10.11) \qquad\qquad \phi(s+t) = \phi(s) + \phi(t)$$

for all real s, t. Thus $m = \exp i\phi$ is a character of $\mathbf{R}$, and being periodic, has the form $\exp ipt$ for some integer p. This concludes the proof.

The theorem has an interesting reformulation. Let h be an endomorphism of l^1, that is, a homomorphism of l^1 into itself. Thus h is a linear mapping, and $h(\mu*\rho) = h(\mu)*h(\rho)$ for all μ, ρ in l^1. We shall show that h necessarily has a trivial form. One trivial possibility is that $h(\rho) = 0$ for all ρ; we exclude this case.

Every endomorphism is continuous, but this is not obvious. We assume the fact for the moment.

For each integer n let e_n be the sequence defined by $e_n(k)$ $= 1$ for $k = n$, $= 0$ for òther k. Then $e_m * e_n = e_{m+n}$ for all m, n. We have $h(e_0) = e_0$ because e_0 is an identity for the algebra. Let $h(e_1)$ $= \mu$. Then $h(e_n) = \mu^n$ for each n (the exponent denotes convolution), and we have

$$(10.12) \qquad \qquad \|\mu^n\|_1 \leq \|h\| \quad \text{(all } n\text{)}.$$

The theorem we have just proved says that $\mu = we_p$ for some integer p, where w has modulus 1. Hence for any ρ in l^1,

$$(10.13) \qquad h(\rho) = \sum \rho(j)\mu^j = \sum \rho(j)w^j e_p^j.$$

It is obvious, conversely, that (10.13) defines a homomorphism for any integer p and constant w of modulus 1.

The problem takes a new form if we think of h as an endomorphism of $\mathbf{A}$, the algebra (under pointwise multiplication) of all absolutely convergent Fourier series. Denote $h(\chi)$ by m, a function of modulus 1 (whose Fourier coefficients are the numbers $\mu(j)$). The *first* equality of (10.13) holds because powers of μ are bounded, even if we do not know the theorem about measures with bounded powers. For any g in $\mathbf{A}$ this equality translates to

$$(10.14) \qquad h(g) = \sum a_j(g)m^j = g \circ m.$$

That is, h carries g to its composition with m. Conversely, if m is any mapping of $\mathbf{T}$ into itself such that $g \circ m$ is in $\mathbf{A}$ whenever g is in $\mathbf{A}$, then the mapping so defined is obviously an endomorphism of l^1. Thus our theorem has this restatement: *if m is any mapping* *of* **T** *into itself such that* $g \circ m$ *belongs to* **A** *for every* g *in* **A**, *then* $m(e^{ix}) = we^{ipx}$ *for some integer* p *and number* w *of modulus* 1. This result emphasizes that no condition of smoothness can characterize the functions of **A**.

The fact that every endomorphism is continuous belongs to Gelfand's theory. For completeness we reproduce the simple and beautiful argument.

It is convenient to give **A** the norm inherited from $l^1(\mathbf{T})$, and study **A** instead of $l^1(\mathbf{T})$.

Let F be a homomorphism of **A** into the complex numbers. That is, F is a linear functional (not assumed to be continuous) such that $F(fg) = F(f)F(g)$. An example of such a functional is evaluation at a point of **T**: $F(f) = f(w)$ for some point w. Our first objective is to show that every homomorphism, not everywhere 0, has this form. For this, we first prove that F is continuous.

The kernel of F (as of any linear functional) is either closed (and then F is continuous) or dense. If we find an open ball in **A** disjoint from the kernel, then the kernel is not dense, and F is continuous. The kernel of F cannot contain any invertible element of **A**, because $1 = F(1) = F(ff^{-1}) = F(f)F(f^{-1})$. (We have assumed that F is not the null functional.) If there is a ball consisting entirely of invertible elements, then this ball is disjoint from the kernel and the kernel is not dense.

Let $\|f\| < 1$ (the norm refers to **A**). Then the series

(10.15) $$1 + f + f^2 + \dots$$

converges in norm, and it is easy to verify that its sum is an inverse for $(1-f)$. Thus each element in the open ball of unit radius about the element 1 of **A** has an inverse. This completes the proof that F is continuous.

$F(\chi)$ is a complex number w, and $F(\chi - w) = 0$ (because $F(1) = 1$). Therefore $\chi - w$ is not invertible; and $|w| = 1$ (Problem 3 below). Thus $F(f) = f(w)$ if f is χ. The same follows immediately if f is a trigonometric polynomial, and since F is continuous, the formula holds for all f in **A**. This proves, as promised above, that every multiplicative linear functional on **A** is evaluation at some point of **T** (except for the null functional).

Now let h be a nonnull endomorphism of **A**. Given w in **T**, define $F(f) = (hf)(w)$. We verify that F is a multiplicative linear functional. F is not the null functional because $F(1) = 1$. Therefore F must be evaluation at another point w' of **T**. If m denotes the function that carries w to w', we have $h(f) = f \circ m$. (We showed above that a *continuous* endomorphism has this form; now this is proved without assuming that h is continuous.)

Finally, the continuity of h follows from the closed graph theorem. For if (f_n) converges to f in **A**, and if $f_n \circ m$ converges to a function g, then $h(f) = g$ because convergence in **A** implies pointwise convergence.

Problems

1. Prove Lemma 1.

2. Show that the series (10.15) gives an inverse for $1-f$ as asserted.

3. Show that $\chi - w$ is invertible in **A** if $|w| \neq 1$.

Chapter 2
The Fourier Integral

1. Introduction

The Fourier integral was introduced in Sections 2 and 3 of Chapter 1, and some results were proved analogous to those already known for Fourier series. Now the Fourier integral is our subject. First the things we know will be summarized.

The Fourier transform $\hat{f}$ of a function f in $\mathbf{L}^1(\mathbf{R})$, defined by (2.1) of Chapter 1, is a bounded continuous function on $\mathbf{R}$. The Riemann-Lebesgue lemma asserts that it tends to 0 at $\pm\infty$.

The convolution of two summable functions is defined almost everywhere, the convolution is summable, and

$$(1.1) \qquad \|f*g\|_1 \leq \|f\|_1 \|g\|_1.$$

For summable functions f, g we have $(f*g)\hat{} = \hat{f}\hat{g}$. If f and g are in $\mathbf{L}^2(\mathbf{R})$, $f*g$ is defined everywhere, and is a bounded continuous function.

An approximate identity on $\mathbf{R}$ (Problem 8 of Chapter 1, Section 3) is a family of functions (e_n) with these properties: each e_n is non-negative, has integral 1, and for each positive ϵ satisfies

$$(1.2) \qquad \lim_{n\to\infty} \int_{-\epsilon}^{\epsilon} e_n(x)\, dx = 1.$$

For any f in $\mathbf{C_0}(\mathbf{R})$ (the space of continuous functions tending to 0 at $\pm\infty$), e_n*f converges to f uniformly. If f is in $\mathbf{L}^p(\mathbf{R})$ with p finite, the convergence is in $\mathbf{L}^p(\mathbf{R})$; if f is bounded, it is in the weak$*$ topology of $\mathbf{L}^\infty(\mathbf{R})$ as the dual of $\mathbf{L}^1(\mathbf{R})$.

As on the circle, a convolution of two functions is smooth if one of the factors is. A useful specific result of this kind is: *if f has compact support and a continuous derivative, and g is in* $\mathbf{L}^1(\mathbf{R})$*, then f∗g has continuous derivative.* This fact, with Fejér's theorem, shows that every g in $\mathbf{L}^1(\mathbf{R})$ can be approximated by functions with continuous derivative, for it is easy to construct an approximate identity having the properties of f.

The *inverse Fourier transform* is defined as

$$(1.3) \qquad\qquad F(x) = \frac{1}{2\pi} \int\limits_{-\infty}^{\infty} f(y)\, e^{ixy}\, dy$$

whenever the formula makes sense. The corollary of Theorem $1'$ (Section 2 of Chapter 1) asserted that if f is summable, then f is the inverse transform of $\hat{f}$ at any point where f satisfies a Lipschitz condition. However $\hat{f}$ need not be summable, so the integral in (1.3) has to be a limit of integrals from $-A$ to B as A, B tend to ∞.

The inversion theorem enables us to prove a unicity theorem, as for Fourier series.

Theorem 7. *If f is in* $\mathbf{L}^1(\mathbf{R})$ *and* $\hat{f} = 0$ *everywhere, then* $f = 0$.

Let (e_n) be an approximate identity consisting of continuously differentiable functions with compact support. Then $(e_n∗f)\hat{} = \hat{e}_n\hat{f} = 0$. Since $e_n∗f$ is continuously differentiable, the inversion theorem implies that it vanishes. But the convolution tends to f in $\mathbf{L}^1(\mathbf{R})$, so $f = 0$.

Problems

1. State and prove a localization theorem for $\mathbf{L}^1(\mathbf{R})$.

2. (a) Define the Fourier-Stieltjes transform of a bounded complex Borel measure on the line. (b) Show that the transform is a continuous function. (c) Define the convolution of a measure with a function in $C_0(R)$. Define the convolution of two measures. (d) Show that if μ is a measure and (e_n) an approximate identity on the line, then $e_n*\mu$ tends to μ in the weak$*$ topology of measures.

3. Construct an approximate identity on R consisting of functions with compact support and continuous derivative.

4. The set of all Fourier transforms of functions in $L^1(R)$ is contained in $C_0(R)$. Show that it is dense. [The dual of $C_0(R)$ is the space of finite complex Borel measures on the line.]

5. Suppose that (μ_n) is a sequence of measures converging to μ in the weak$*$ topology, and such that $\hat{\mu}_n$ converges pointwise to a *continuous* function h. Show that $\hat{\mu} = h$. [The difficulty is that exponentials are not in $C_0(R)$.]

6. Prove the unicity theorem for Fourier-Stieltjes transforms in three ways. (a) Show that $\mu*\hat{f} = 0$ if f is in $L^1(R)$ and $\hat{\mu} = 0$. Prove directly that such functions $\hat{f}$ are dense in $C_0(R)$. (b) For any real t, define two analytic functions:

$$F(z) = \int_{-\infty}^{t} e^{-ixz}\, d\mu(x), \qquad G(z) = -\int_{t}^{\infty} e^{-ixz}\, d\mu(x)$$

in the upper and lower halfplanes, respectively. Each function is continuous on the closure of its halfplane. The hypothesis that $\hat{\mu} = 0$ means that $F = G$ on the real axis. Use Morera's theorem and Liouville's theorem to show that F and G are 0. Then by varying t, show that μ is 0. (This proof is due to D. J. Newman.) (c) Fix a positive number t. Define a measure ν on $[0, t)$ by setting

$$\nu(E) = \sum_{k} \mu(E + kt).$$

Show that ν has Fourier-Stieltjes coefficients (relative to the given interval) all 0. Vary t and deduce that μ is null.

2. Kernels on R

In this section we define the Dirichlet, Fejér and Poisson kernels on $\mathbf{R}$, and show that they have properties like those of the corresponding kernels on $\mathbf{T}$.

The kernels will be defined by telling what their Fourier transforms are:

$$(2.1) \qquad \hat{D}_t(y) = 1 \quad (|y| \leq t), \quad = 0 \quad (|y| > t)$$

$$(2.2) \qquad \hat{K}_t(y) = 1 - \frac{|y|}{t} \quad (|y| \leq t), \quad = 0 \quad (|y| > t)$$

$$(2.3) \qquad \hat{P}_u(y) = e^{-u|y|}.$$

The continuous parameters t, u are positive; t tends to ∞, u tends to 0. $\hat{K}_t$ is a triangular function of height 1 centered at the origin.

Taking the inverse Fourier transform of (2.1) gives

$$(2.4) \qquad D_t(x) = \frac{\sin tx}{\pi x}.$$

This is not a summable function, so it does not have a Fourier transform in the ordinary sense. However, $\hat{D}_t$ is a summable function to which the corollary of Theorem 1′ (Section 2 of Chapter 1) can be applied:

$$(2.5) \qquad \hat{D}_t(y) = \lim_{A, B \to \infty} \int_{-A}^{B} D_t(x)\, e^{-ixy}\, dx \quad (y \neq \pm t).$$

Of course D_t could not be summable, because its transform is not continuous.

We could calculate the inverse transform of $\hat{K}_t$ with only a little more trouble, but even that is unnecessary. From the definitions we have

$$(2.6) \qquad 2t\hat{K}_{2t} = \hat{D}_t * \hat{D}_t.$$

(It is easy to convolve $\hat{D}_t$ with itself.) Both sides are bounded functions with compact support. The inverse Fourier transform, like the transform itself, carries convolution to multiplication, with the complication of a factor 2π. Thus (2.6) leads to

$$(2.7) \qquad 2tK_{2t}(x) = 2\pi D_t(x)^2, \quad \text{or}$$

$$(2.8) \qquad K_t(x) = \frac{1}{2\pi t}\Big(\frac{\sin tx/2}{x/2}\Big)^2.$$

This function is positive and summable. Its transform is the function (2.2), by the inversion theorem. Its integral is 1 because its transform equals 1 at 0. For any positive ϵ,

$$(2.9) \qquad \int_{|x|>\epsilon} K_t(x)\,dx \leq \frac{1}{2\pi t}\int_{|x|>\epsilon}\frac{4}{x^2}\,dx,$$

which tends to 0 as t tends to ∞. Hence (K_t) is an approximate identity on **R**.

An easy computation gives for the inverse Fourier transform of (2.3)

$$(2.10) \qquad P_u(x) = \frac{u}{\pi(u^2+x^2)}.$$

Thus P_u is positive, and we can check that

$$(2.11) \qquad \lim_{u \downarrow 0} \int_{-\epsilon}^{\epsilon} P_u(x)\,dx = 1$$

for each positive ϵ. Hence (P_u) is an approximate identity, with u directed downwards to 0.

Here is another useful inversion theorem.

Theorem 8. *If f and $\hat{f}$ are both summable, then f is the inverse Fourier transform of $\hat{f}$ at each point.*

The conclusion is not stated quite accurately. We mean that f is almost everywhere equal to a continuous function, which is equal at every point to the inverse transform of $\hat{f}$.

For positive u, $P_u * f$ is differentiable, so the inversion theorem gives

$$(2.12) \qquad P_u * f(x) = \frac{1}{2\pi} \int_{-\infty}^{\infty} e^{-u|y|}\, \hat{f}(y)\, e^{ixy}\, dy.$$

(This can be verified directly by inserting the integral defining $\hat{f}$ on the right side, then changing the order of integration.) As u decreases to 0, the right side tends to the inverse transform of $\hat{f}$ at x. On the left, $P_u * f(x)$ tends to f a.e. at least on a subsequence of u, and this proves the theorem.

The kernel P_u belongs to $\mathbf{L}^q(\mathbf{R})$ for every q. Hence for f in $\mathbf{L}^p(\mathbf{R})$ (where p and q are conjugate exponents) we can define

$$(2.13) \qquad F(x + iu) = P_u * f(x) = \frac{1}{\pi} \int_{-\infty}^{\infty} \frac{u\,f(s)\,ds}{u^2 + (x - s)^2}.$$

This is the *Poisson integral* of f, giving a harmonic extension of f to the upper halfplane. The fact that F is harmonic can be verified by differentiation under the integral sign. If $f(s)\,ds$ is replaced by a measure $d\mu(s)$ such that $d\mu(s)/(1 + s^2)$ is finite, the function F is still defined and harmonic.

The Poisson kernel on the line has a semigroup property, as on the circle: $P_u*P_v = P_{u+v}$ for all positive u, v. This is obvious from (2.3). It follows that

$$(2.14) \qquad \int_{-\infty}^{\infty} |F(x + iu)|^p \, dx$$

decreases as u increases, for any finite p and f in $\mathbf{L}^p(\mathbf{R})$. If f is bounded, the upper bound of $|P_u*f|$ is a decreasing function of u.

Let F be a function harmonic in the upper halfplane. Define $f_u(x) = F(x + iu)$ for positive u. We ask whether there is a function f in some Lebesgue space such that $f_u = P_u*f$ for positive u.

The answer is like that on the circle. If f_u belongs to $\mathbf{L}^p(\mathbf{R})$ for each u, with norm bounded by a constant independent of u, then there is such a boundary function f in $\mathbf{L}^p(\mathbf{R})$ (if $p > 1$), or there is a boundary measure μ (if $p = 1$).

The proof is analogous to that on the circle, but there is a complication. Every harmonic function on the disk has a series representation ((6.4) of Chapter 1); this follows from the fact that a real harmonic function is the real part of an analytic function, which is the sum of a Taylor series. From the series it was obvious that $P_r*f_s = f_{rs}$ for r, s between 0 and 1. The corresponding statement now is that $P_u*f_v = f_{u+v}$ for positive u and v, and this is not so obvious. We shall prove this fact, after which the rest is easy.

Lemma. *Let F be bounded and continuous on the closed upper halfplane, and harmonic on the open halfplane. If $F(x) = 0$*

for all real x, then F = 0.

A conformal map of the upper halfplane onto the unit disk carries F to a function that is continuous on the closed disk except perhaps for the point that was the image of ∞, null on the boundary except at that point, harmonic and bounded on the open disk. This harmonic function is the Poisson integral of its boundary values and therefore is 0 everywhere. This proves the lemma.

Now let F be a function harmonic in the upper halfplane with the associated family (f_u) bounded in $\mathbf{L}^p(\mathbf{R})$, $1 \leq p \leq \infty$. For any function g in $\mathbf{L}^q(\mathbf{R})$, where q is the exponent conjugate to p, $(g*f_u)$ is a family of bounded continuous functions, with norms in $\mathbf{L}^\infty(\mathbf{R})$ bounded for $u > 0$. If we take g continuous with compact support, it is easy to prove that $G(x+iu) = g*f_u(x)$ is harmonic (Problem 4 below). Fix $v > 0$. Then $G(x+iu+iv) = g*f_{u+v}(x)$ is harmonic and bounded for $u > 0$, and continuous for $u \geq 0$. The same is true for the function $H(x+iu) = P_u*g*f_v$. (The continuity of H at the boundary is not quite guaranteed by Fejér's theorem, because $g*f_v$ may not belong to $\mathbf{C}_0(\mathbf{R})$, but follows easily from it.) By the lemma, $G(x+iu+iv) = H(x+iu)$ for $u \geq 0$ and all x.

This result states that P_u*f_v defines the same linear functional as f_{u+v} on the set of functions g that are continuous with compact support. This set is dense in $\mathbf{L}^p(\mathbf{R})$ for $p < \infty$, and *-dense in $\mathbf{L}^\infty(\mathbf{R})$. Hence P_u*f_v equals f_{u+v} a.e. Since both functions are continuous, they are equal. This is what we wanted to show.

Finally we complete the proof that (f_u) is the family (P_u*f) for a function f in $\mathbf{L}^p(\mathbf{R})$ (or a measure μ if $p = 1$). By compactness of balls in the dual space, there is a function f or a measure μ such that every $*$-neighborhood of f or of μ contains f_u for arbitrarily small positive u. Each function P_u belongs to the predual space; hence for each real x and fixed u, $P_u*f(x)$ (or $P_u*\mu(x)$) $= \lim P_u*f_{v_j}(x) = \lim f_{u+v_j}(x) = f_u(x)$ for some sequence (v_j) decreasing to 0. This completes the proof.

It follows, as on the circle, that P_u*f converges to f in the norm of $\mathbf{L}^p(\mathbf{R})$ if $1 \leq p < \infty$, or in the $*$-topology if $p = \infty$; and if μ is a finite complex measure, $P_u*\mu$ tends to μ in the $*$-topology of measures.

Problems

1. Show that $(-xP_u')$ is an approximate identity on $\mathbf{R}$. Use this fact to prove that P_u*f tends to f a.e. if f is in $\mathbf{L}^p(\mathbf{R})$ for some p.

2. Show that functions in $\mathbf{L}^1(\mathbf{R})$ whose Fourier transforms have compact support are dense in the space.

3. Show that if f is in $\mathbf{L}^1(\mathbf{R})$ and $\hat{f}$ has compact support, then f is a.e. equal to the restriction of an entire function to the real axis.

4. Show that if g is continuous with compact support, and F is harmonic in the upper halfplane, then $H(x + iu) = g*f_u(x)$ is harmonic in the halfplane. [Riemann sums for the convolution integral converge uniformly on compact subsets of the halfplane. Or else differentiate under the integral sign.]

5. *Multipliers of* $\mathbf{L}^1(\mathbf{R})$. If k is a Fourier-Stieltjes transform, then $k\hat{f}$ is the transform of a summable function whenever $\hat{f}$ is (namely, of $\mu*f$, where $\hat{\mu} = k$). The converse is a famous theorem of Fekete: *if k multiplies each Fourier transform into a Fourier transform, then k is a Fourier-Stieltjes transform.* Prove this by

following these steps. (a) Show that k is continuous. (b) The operation in $\mathbf{L}^1(\mathbf{R})$ defined by $(Tf)^\hat{} = k\hat{f}$ is continuous. [Use the closed graph theorem.] (c) Let (e_n) be an approximate identity. Find a weak* limit for (Te_n) in $\mathbf{M}(\mathbf{R})$, and show that its Fourier-Stieltjes transform is k. [Use Problem 5 of the last section.]

6. Evaluate the integrals

$$\int_{-\infty}^{\infty} \frac{\sin x}{x}\, dx, \qquad \int_{-\infty}^{\infty} \left(\frac{\sin x}{x}\right)^2 dx.$$

3. The Plancherel theorem

The Parseval relation is the statement that the Fourier transform is an isometry from $\mathbf{L}^2(\mathbf{T})$ onto $\mathbf{l}^2$, when the first Lebesgue space is based on normalized Lebesgue measure and the second space on the measure that gives each point of $\mathbf{Z}$ unit mass. The *Plancherel theorem* says that the Fourier transform is an isometry of $\mathbf{L}^2(\mathbf{R})$ onto itself, when the space is based on the measure $dx/\sqrt{2\pi}$. There is a qualification: the Fourier transform is defined on $\mathbf{L}^1(\mathbf{R})$; but on the line, $\mathbf{L}^2$ is not contained in $\mathbf{L}^1$. Therefore the Fourier transform has to be extended to the rest of $\mathbf{L}^2(\mathbf{R})$. The Plancherel theorem consists of several statements that together describe this isometry.

We introduce a factor in the definition of the Fourier transform, and call it the Plancherel transform:

$$(3.1) \qquad F(y) = (2\pi)^{-1/2} \int_{-\infty}^{\infty} f(x)\, e^{-ixy}\, dx.$$

(We shall reserve the notation $\hat{f}$ for the transform of f without the factor.)

Lemma. *There is a dense subspace of $\mathbf{L}^2(\mathbf{R})$ contained in*

$\mathbf{L}^1(\mathbf{R})$ *on which the Plancherel transform is isometric.*

Define $\tilde{f}(x) = \bar{f}(-x)$. Then $\hat{\tilde{f}} = \overline{\hat{f}}$. If f is differentiable with compact support, the same is true of $f * \tilde{f}$. The transform of $f * \tilde{f}$ is $|\hat{f}|^2$, and $f * \tilde{f}(0) = \|f\|_2^2$. By Theorem 1′ of Chapter 1, for such functions we have

$$(3.2) \qquad f * \tilde{f}(0) = \lim_{A, B \to \infty} (2\pi)^{-1} \int_{-A}^{B} |\hat{f}(y)|^2 \, dy.$$

This shows that $\hat{f}$ is square-summable. Replacing $\hat{f}$ by $(2\pi)^{1/2} F$ (see (3.1)) gives

$$(3.3) \qquad \int_{-\infty}^{\infty} |f(x)|^2 \, dx = \int_{-\infty}^{\infty} |F(y)|^2 \, dy.$$

The functions f for which this is shown form a dense subspace of $\mathbf{L}^2(\mathbf{R})$, so the lemma is proved.

Denote by Q the isometry that maps f to F. Since Q is defined on a dense subspace of $\mathbf{L}^2(\mathbf{R})$ it has a unique continuous extension to a linear isometry of all of $\mathbf{L}^2(\mathbf{R})$ into itself. We call this extension Q again. The range of Q is either the whole space or a closed proper subspace. If we show that the range is dense, then it is actually all of $\mathbf{L}^2(\mathbf{R})$.

The adjoint Q^* of Q is the operator defined by the relation

$$(3.4) \qquad (Qf, h) = (f, Q^*h)$$

for all f, h in the space. A simple computation (Problem 4 below) shows that Q^* is essentially the Fourier transform again, and therefore is an isometry when properly normalized, by what has just been proved. Thus its null space is trivial. The null space of

Q^* is the complement of the range of Q. Hence the range of Q is dense, as we wished to show.

The operator Q, which is the normalized Fourier transform on functions that are differentiable and compactly supported, has been shown to be isometric and to have a unique continuous extension to a unitary operator on $\mathbf{L}^2(\mathbf{R})$. This operator is the *Plancherel transform*. Its adjoint is its inverse, and this is formally the inverse Fourier transform, except that the factor $(2\pi)^{-1}$ in formula (2.7) of Chapter 1 is replaced by $(2\pi)^{-1/2}$.

For any h in $\mathbf{L}^2(\mathbf{R})$, the functions h_n equal to h on $(-n,\,n)$ and to zero elsewhere are summable, and converge to h in $\mathbf{L}^2(\mathbf{R})$. Hence their transforms tend in norm to the transform of h:

$$(3.5) \qquad H(y) = \mathrm{l.i.m.}_n \; (2\pi)^{-1/2} \int_{-n}^{n} h(x)\, e^{-ixy}\, dx.$$

The curious notation "l.i.m." is a pun of Norbert Wiener; it means "limit in mean." The notation is inaccurate, because the l.i.m. is the limit of functions in the norm of $\mathbf{L}^2(\mathbf{R})$, not a limit at the point y. However if the pointwise limit exists a.e. even for a subsequence of n, the limit function is the Plancherel transform of h, because when a sequence of functions converges in a Lebesgue space, a subsequence converges a.e. to the limit function.

The Plancherel transform carries elements of $\mathbf{L}^2(\mathbf{R})$ to elements of the same space, which are equivalence classes of functions. That is, the transform H is only determined up to sets of measure 0. This is different from the Fourier transform of a function in $\mathbf{L}^1(\mathbf{R})$, which is a continuous function, defined everywhere. Even though the Plancherel transform is defined as an extension of the Fourier transform, it has a different character.

Nevertheless, we use the notation $\hat{f}$ for functions f in $\mathbf{L}^2(\mathbf{R})$, to denote the Plancherel transform of f multiplied by $\sqrt{2\pi}$.

Problems

1. Denote by $\mathbf{L}$ the linear space of all functions $f + g$ where f is in $\mathbf{L}^1(\mathbf{R})$ and g in $\mathbf{L}^2(\mathbf{R})$. Invent a definition of Plancherel transform for functions in $\mathbf{L}$. Show that $\mathbf{L}^p(\mathbf{R})$ is contained in $\mathbf{L}$ for $1 < p < 2$. Prove the unicity theorem in $\mathbf{L}^p(\mathbf{R})$.

2. Show that if f and g are in $\mathbf{L}^2(\mathbf{R})$, so that fg is summable, then $2\pi(fg)^\smallfrown = \hat{f} * \hat{g}$ (where the convolution makes sense because $\hat{f}$ and $\hat{g}$ are square-summable). [Note that both sides are continuous functions.]

3. It was stated, but not strictly proved, that the two transforms coincide on $\mathbf{L}^1(\mathbf{R}) \cap \mathbf{L}^2(\mathbf{R})$ except for a constant factor. Show that this is so. [The Plancherel transform was defined only for differentiable functions with compact support to begin with.]

4. Calculate the adjoint of the isometry Q.

4. Another convergence theorem; the Poisson summation formula

The theorem on convergence of Fourier series in Chapter 1 was based on Mercer's theorem. Here is another convergence theorem, of different character.

Theorem 9. *Let f be a function of bounded variation on $\mathbf{T}$, normalized so that $f(e^{ix})$ is the average of its limits from left and right at each point. Then f is the sum of its Fourier series at each point.*

This result, with the principle of localization, shows that a function of $\mathbf{L}^1(\mathbf{T})$ is the sum of its Fourier series at each point in the interior of an interval in which it is of bounded variation and normalized as above.

Lemma 1. *There is a number K such that*

$$(4.1) \qquad \left| \int_a^b D_n \, d\sigma \right| \leq K$$

for all a, b such that $|b - a| \leq 2\pi$, and all positive integers n.

The proof is asked for in Problem 6 below.

Lemma 2. *Let $g(e^{ix}) = -1$ for x in $(-\pi, 0)$, $= 1$ on $(0, \pi)$. Then*

$$(4.2) \qquad \lim_{n \to \infty} \sum_{-n}^{n} a_k(g) = 0.$$

Each sum in (4.2) actually equals 0, and there is nothing to prove. [Recall example 4(a) at the end of Chapter 1, Section 1.]

The lemma is a special case of the theorem. By adding a multiple of a translate of g to f, we can make f continuous at any point we choose. Thus Lemma 2 reduces the proof of the theorem to the case of convergence at a point of continuity.

Lemma 3. *If h is the characteristic function of a subinterval (a, b) of $[-\pi, \pi]$, normalized at the endpoints, then the Fourier series of h converges boundedly to h at each point.*

Convergence at points of continuity of h is assured by our old convergence theorem, and at the jump points, by Lemma 2. Boundedness follows from Lemma 1, because the partial sums of the Fourier series of h are

$$(4.3) \qquad D_n * h(e^{ix}) = \int_a^b D_n(e^{i(x-t)}) \, d\sigma(t).$$

Lemma 4. *Let μ be any finite measure and h the function of Lemma 3. Then $\mu * h$ is the sum of its Fourier series at every point.*

The convolution is meant in the literal sense of (4.4) of Chapter 1, even though h is not continuous. The usual computation shows that the Fourier coefficients of $\mu * h$ are $a_k(\mu) a_k(h)$. The symmetric partial sums of the Fourier series of $\mu * h$ are

$$(4.4) \qquad \sum_{-n}^{n} a_k(\mu)\, a_k(h)\, e^{ikx} = \int \sum_{-n}^{n} a_k(h)\, e^{ik(x-t)}\, d\mu(t).$$

The sum on the right converges boundedly to $h(e^{i(x-t)})$ by Lemma 3. The dominated convergence theorem shows that the right side converges to $\mu * h(e^{ix})$.

Now we can finish the proof of the theorem. Let f be a normalized function of bounded variation, and μ the associated measure: $\mu(a, b) = f(e^{ib}) - f(e^{ia})$ if a, b are points of continuity of f in $[0, 2\pi)$, and μ has point masses corresponding to the jumps of f. Without loss of generality we investigate the convergence of the Fourier series of f at 1, and we assume that f is continuous there, and vanishes outside $(-\pi/3, \pi/3)$. Let $h(e^{ix})$ be the characteristic function of $(0, \pi/2)$. Then $\mu * h = f$ on a neighborhood of 1, at least where f is continuous. We have shown that $\mu * h$ is the sum of its Fourier series; hence the same is true for f at points of continuity near 1, as was to be proved.

The proof shows that the Fourier series converges to the average of left and right limiting values at points where f has a simple discontinuity.

We come to the Poisson summation formula. Let f be a summable function on **R**. Then

(4.5) $\sum\limits_{-\infty}^{\infty} \int\limits_{0}^{2\pi} |f(x+2\pi n)|\,dx \;=\; \int\limits_{-\infty}^{\infty} |f(x)|\,dx < \infty.$

Thus the sum

(4.6) $f^*(e^{ix}) = \sum\limits_{-\infty}^{\infty} f(x+2\pi n)$

converges absolutely a.e. and defines a summable function on $\mathbf{T}$. The Fourier coefficients of this function are

(4.7) $a_n(f^*) = \int f^*(e^{ix})\,e^{-nix}\,d\sigma(x) = \dfrac{1}{2\pi} \int\limits_{-\infty}^{\infty} f(x)\,e^{-nix}\,dx = \dfrac{1}{2\pi}\hat{f}(n).$

Thus f^* has Fourier series

(4.8) $f^*(e^{ix}) \sim \dfrac{1}{2\pi} \sum \hat{f}(n)\,e^{nix}.$

If f^* is the sum of its Fourier series at $x = 0$,

(4.9) $\sum f(2\pi n) \;=\; \lim\limits_{n\to\infty} \dfrac{1}{2\pi} \sum\limits_{-n}^{n} \hat{f}(k).$

This is the Poisson summation formula.

  No single hypothesis gives the best condition for the truth of (4.9), because there is no best theorem about the convergence of Fourier series. The simplest useful condition is given by the theorem we have just proved: *the formula holds under the hypothesis that f is summable, of bounded variation on* $\mathbf{R}$, *and normalized.* For then (Problem 3 below) (4.6) converges everywhere, f^* is a normalized function of bounded variation on $\mathbf{T}$, and (4.9) says that f^* is the sum of its Fourier series at 1.

  The point 1 is not special, and the formula can be written

more generally

(4.10) $$\sum f(x+2\pi n) = \lim_{n\to\infty} \frac{1}{2\pi} \sum_{-n}^{n} \hat{f}(k)\, e^{kix}.$$

This holds for all x if the hypotheses just mentioned are satisfied.

Problems

1. Prove the famous theta relation: if z is complex and $\Re z > 0$, then

$$\sum_{-\infty}^{\infty} e^{-n^2\pi z} = z^{-1/2} \sum_{-\infty}^{\infty} e^{-k^2\pi/z}.$$

[Recall Problem 7 of Chapter 1, Section 2.]

2. Evaluate $\sum_{1}^{\infty} 1/(n^2+1)$ by means of the Poisson formula. [This sum was evaluated in Problem 7 of Chapter 1, Section 4.]

3. Let $V(f)$ denote the total variation of f on $\mathbf{T}$. Show that $V(f+g) \leq V(f) + V(g)$. Show that if (f_n) is a sequence of functions of bounded variation on $\mathbf{T}$ with $\sum V(f_n) < \infty$, and if $f_n(1) = 0$ for each n, then $\sum f_n$ converges absolutely and uniformly, and the sum is of bounded variation. If each f_n is normalized, then the sum is normalized. Show that (4.6) converges absolutely and uniformly if f is summable and of bounded variation on the line.

4. Obtain a formula expressing $\sum f(nu)$ in terms of $\hat{f}$, where u is an arbitrary positive number.

5. Show that if f is a function of bounded variation on $\mathbf{T}$, then $a_n(f) = O(1/n)$.

6. Prove Lemma 1.

5. Bochner's theorem

A complex function f on the line is *positive definite* if

(5.1) $$\sum f(x_j - x_k)\, c_j \overline{c}_k \geq 0$$

for every choice of finitely many complex numbers (c_j) and real numbers (x_j). This definition is algebraic and has only algebraic consequences. However as soon as f is measurable, an analogue of Herglotz' theorem (Section 8 of Chapter 1) is valid.

Bochner's theorem. *If f is positive definite and measurable, then for some positive measure μ on the line we have $f = \hat{\mu}$ almost everywhere.*

As in Chapter 1 on the circle, it is easy to prove that the Fourier-Stieltjes transform of a positive measure is positive definite (and continuous). Now we assert conversely that a continuous positive definite function is the transform of a positive measure, and further that every measurable positive definite function is a.e. equal to a continuous one.

First we shall prove a somewhat different statement. Say that a measurable function f on the line is *positive definite in the integral sense* if f is bounded and

(5.2) $$\int\int f(x - y)\, g(x)\, \overline{g}(y)\, dx\, dy \geq 0$$

for every summable function g. (The integral exists absolutely because f is bounded.) Unlike (5.1), this condition applies to functions defined only modulo Lebesgue null sets. We shall show that *every function positive definite in the integral sense is equal a.e. to $\hat{\mu}$, for some positive measure μ.* Then Bochner's theorem will be proved by showing that a measurable positive definite function is positive definite in the integral sense.

We shall imitate the proof of Herglotz' theorem. Let

$h = h_A$ be the characteristic function of $[0, A]$ divided by $A^{1/2}$, and $\tilde{h}(x) = \overline{h}(-x)$ as usual. (Here h is real, of course.) Then $h * \tilde{h}$ is the triangular function supported on $(-A, A)$, with value 1 at the origin. (This is the transform of the Fejér kernel K_A.) In (5.2) take $g(x) = h(x) \exp itx$, where t is a real variable. After a change of variable we find that

$$(5.2) \qquad k(t) = \int f(x) \, h * \tilde{h}(x) \, e^{itx} \, dx \geq 0.$$

Thus k is the Fourier transform of a summable function and is non-negative. We want to show that k is summable.

Integrate k against the transform of a Fejér kernel:

$$(5.3) \qquad \int k(t) \, \hat{K}_B(t) \, dt = \int \int f(x) \, h * \tilde{h}(x) \, \hat{K}_B(t) \, e^{itx} \, dt \, dx.$$

The integrand on the right is continuous with bounded support, so Fubini's theorem applies to show that this equals

$$(5.4) \qquad 2\pi \int f(x) \, h * \tilde{h}(x) \, K_B(x) \, dt.$$

This quantity is at most equal to $\|f\|_\infty$, because (K_B) is an approximage identity, and $h * \tilde{h}(x)$ is bounded by 1. But the left side of (5.3) increases to the integral of k. Hence k is summable, with norm at most $\|f\|_\infty$.

The function k depends on the parameter A, but this bound is independent of A. Therefore the family of measures $(k_A \, d\sigma)$ has a weak* accumulation point μ. By Theorem 8 of Section 2,

$$(5.5) \qquad \frac{1}{2\pi} \int k_A(t)\, e^{-itx}\, dt = f(x)\, h_A * \tilde{h}_A(x) \quad \text{a.e.}$$

This shows that f is a.e. equal to a continuous function. As A tends to ∞ the right side tends boundedly to f. If the exponential on the left side were in the space $\mathbf{C_0(R)}$ of continuous functions tending to 0 at $\pm\infty$, it would follow immediately that $2\pi f = \hat{\mu}$.

But this is not so, and another step is needed. Multiply (5.5) by any summable function $g(t)$ and integrate:

$$(5.6) \qquad \frac{1}{2\pi} \int k_A(t)\, \hat{g}(t)\, dt = \int f(x)\, h_A * \tilde{h}_A(x)\, g(x)\, dx.$$

Since $\hat{g}$ is in $\mathbf{C_0(R)}$, we can take the limit in A to obtain

$$(5.7) \qquad \frac{1}{2\pi} \int \hat{g}(t)\, d\mu(t) = \int f(x)\, g(x)\, dx.$$

Now replacing $\hat{g}$ by the integral that defines it shows that $2\pi f = \hat{\mu}$. Furthermore μ as a linear functional on $\mathbf{C_0(R)}$ is non-negative on positive functions, and it follows that μ is a positive measure.

Finally, we must show that a measurable positive definite function f is positive definite in the integral sense. This result and the following ingenious proof are due to F. Riesz.

As in the case of positive definite sequences, it follows from the definition (5.1) that f is bounded. Let g be any Borel function that is summable and square-summable. In (5.1), choose the numbers c_j to be the values $g(x_j)$:

$$(5.8) \qquad \sum_{j,\,k=1}^{m} f(x_j - x_k)\, g(x_j)\, \overline{g}(x_k) \geq 0.$$

The numbers x_j are arbitrary; regard them as variables and integrate with respect to each of them from $-A$ to A. The m terms with $j = k$ give

$$(5.9) \qquad m(2A)^{m-1} f(0) \int_{-A}^{A} |g(x)|^2 \, dx.$$

There are $m^2 - m$ other terms, all equal:

$$(5.10) \qquad (m^2 - m)(2A)^{m-2} \int_{-A}^{A} \int_{-A}^{A} f(x - y) \, g(x) \, \overline{g}(y) \, dx \, dy.$$

The sum of these terms is non-negative.

Add the terms and divide by $(m^2 - m)(2A)^{m-2}$:

$$(5.11) \quad \frac{m}{m^2 - m} \, 2A f(0) \int_{-A}^{A} |g(x)|^2 \, dx + \int_{-A}^{A} \int_{-A}^{A} f(x - y) \, g(x) \, \overline{g}(y) \, dx \, dy \geq 0.$$

Letting m tend to ∞ shows that

$$(5.12) \qquad \int_{-A}^{A} \int_{-A}^{A} f(x - y) \, g(x) \, \overline{g}(y) \, dx \, dy \geq 0,$$

and since A was arbitrary, (5.2) is proved, at least if g is square-summable as well as summable. By approximation the inequality holds for all summable functions g, and this completes the proof.

Problems

1. Show that the pointwise limit of positive definite functions is positive definite. Show that a positive definite function f is bounded, and satisfies $f(-t) = \overline{f}(t)$ for all real t.

2. Let $f(t) = 1$ when t is an integer, and $= 0$ elsewhere. Verify that f is positive definite.

3. Show that a positive definite function on the line is

continuous if it is continuous at the point 0. [In the definition
(5.1) take three points $0, x, x + u$. By choosing c_1, c_2, c_3 cleverly,
find an estimate for $|f(x + u) - f(x)|$.]

6. The Continuity theorem

Theorem 10. *Let (μ_n) be a sequence of probability measures
on the line whose transforms $(\hat{\mu}_n)$ converge pointwise to a limit
function h. If h is continuous at 0, then $h = \hat{\mu}$ for a probability
measure μ, and μ_n converges to μ in the weak* topology of
measures.*

A *probability measure* is a measure that is positive with
total mass 1. We assume that measures are defined on the Borel
field of the line. This theorem from probability theory is presented
here because it is almost a corollary of Bochner's theorem. Indeed
we needed part of the content of this result to prove Bochner's
theorem in the last section.

Since h is the pointwise limit of positive definite functions,
it is positive definite. By assumption it is continuous at 0 (this
hypothesis is essential); Problem 3 of the last section says that h
is continuous everywhere, so by Bochner's theorem $h = \hat{\mu}$ for some
positive measure μ. This measure μ is a probability measure
because $h(0) = 1$. The only thing left to prove is that μ_n tends to
μ in the weak* topology of measures (that is, in the dual of the
space $\mathbf{C_0(R)}$).

The assumed convergence of the Fourier-Stieltjes transforms means that

$$(6.1) \qquad \int \phi \, d\mu = \lim \int \phi \, d\mu_n$$

for every trigonometric polynomial ϕ, and we want the same convergence for all functions ϕ in $\mathbf{C_0(R)}$.

We use the idea of the last section. Let f be any summable function. Its transform is in $\mathbf{C_0(R)}$, and

$$(6.2) \qquad \int \hat{f} \, d\mu_n = \int \int f(x) \, e^{-itx} \, dx \, d\mu_n(t) = \int f(x) \, \hat{\mu}_n(x) \, dx.$$

The sequence $(\hat{\mu}_n)$ converges boundedly to $\hat{\mu}$, so the right side has limit

$$(6.3) \qquad \int f(x) \, \hat{\mu}(x) \, dx = \int \hat{f} \, d\mu.$$

Thus (μ_n) converges weakly to μ over the subspace of $\mathbf{C_0(R)}$ consisting of all transforms of summable functions. This subspace is dense, and the sequence (μ_n) is bounded, so the sequence converges on all of $\mathbf{C_0(R)}$. This finishes the proof.

Theorem 10 tells less than the whole truth. Denote by $\mathbf{P}$ the set of all Borel probability measures on the line. The elements of $\mathbf{P}$ define linear functionals on all bounded continuous functions by the usual definition:

$$(6.4) \qquad \mu(\phi) = \int \phi \, d\mu.$$

Let $\mathbf{W}$ be any linear space of bounded continuous functions given the uniform norm. Then $\mathbf{P}$ is part of the dual space of $\mathbf{W}$, and has a weak∗ topology with respect to $\mathbf{W}$. The interesting fact is that these topologies on $\mathbf{P}$ are all the same, provided that $\mathbf{W}$ is not too small.

Theorem 11. *Suppose that* $\mathbf{W}$ *contains* $\mathbf{C_0(R)}$. *Then the* *weak* topology of* **P** *with respect to* **W** *is the same as the weak* topology of* **P** *as the dual of* $\mathbf{C_0(R)}$.

Denote by $\mathbf{P_1}$ the set **P** with the weak* topology relative to $\mathbf{C_0(R)}$, and by $\mathbf{P_2}$ the same set with the weak* topology relative to **W**. Let T be the mapping from $\mathbf{P_1}$ to $\mathbf{P_2}$ defined by $T\mu = \mu$. The theorem asserts that T is a homeomorphism.

If a net (μ_α) converges over **W**, then it converges over $\mathbf{C_0(R)}$ *a fortiori* and to the same limit. That is, T^{-1} is continuous. We must prove that T is continuous.

Lemma. *If E is a subset of* $\mathbf{P_1}$, *and μ belongs to the closure of E, then some sequence in E converges to μ.*

This follows from the fact that $\mathbf{C_0(R)}$ is separable and the measures of **P** have norms that are bounded.

Let E be a closed subset of $\mathbf{P_2}$. We must show that $T^{-1}E = E$ is closed in $\mathbf{P_1}$. Let μ be in the $\mathbf{P_1}$-closure of E, and (μ_n) a sequence in E converging $(\mathbf{P_1})$ to μ. Define ϕ_k to be the triangular function equal to 1 at 0, vanishing outside $(-k, k)$, and linear on the intervals $(-k, 0)$ and $(0, k)$. These functions belong to $\mathbf{C_0(R)}$, and tend boundedly to 1 everywhere as k tends to ∞. Hence given $\epsilon > 0$ there is a k such that

$$(6.5) \qquad\qquad \int \phi_k \, d\mu > 1 - \epsilon.$$

It follows that for all sufficiently large n

$$(6.6) \qquad\qquad \int \phi_k \, d\mu_n > 1 - 2\epsilon.$$

Since these are probability measures, (6.5) and (6.6) mean that the interval $(-k, k)$ contains almost all the mass of all the μ_n

$(n \geq n(\epsilon))$ and μ at once. This is the fact that we shall use.

If μ is not in E, there is a $\mathbf{P_2}$-neighborhood of μ disjoint from E. That is, there are functions $\psi_1, \ldots, \psi_r$ in $\mathbf{W}$ and $\eta > 0$ such that no measure ν in E satisfies

$$(6.7) \qquad |\nu(\psi_j) - \mu(\psi_j)| < \eta \qquad (j = 1, \ldots, r).$$

Therefore for one j at least,

$$(6.8) \qquad |\mu_n(\psi_j) - \mu(\psi_j)| \geq \eta$$

for infinitely many n. We fix $\psi = \psi_j$ and pass to a subsequence of (μ_n), in order to obtain a sequence (ν_n) from E convergent $(\mathbf{P_1})$ to μ and a single ψ in $\mathbf{W}$ such that

$$(6.9) \qquad |\nu_n(\psi) - \mu(\psi)| \geq \eta \qquad (\text{all } n).$$

For each k and n,

$$(6.10) \qquad |\mu(\psi) - \nu_n(\psi)| \leq |\mu(\psi) - \mu(\phi_k\psi)| +$$
$$|\mu(\phi_k\psi) - \nu_n(\phi_k\psi)| + |\nu_n(\phi_k\psi) - \nu_n(\psi)|.$$

Let $\epsilon = \eta/6$. By (6.5) and (6.6) we can choose and fix k so large that the first and last terms on the right are at most ϵ for all large n. The middle term tends to 0 as n tends to ∞ because $\phi_k\psi$ is in $\mathbf{C_0(R)}$, so it is smaller than ϵ for all sufficiently large n. Thus the right side is less than $3\epsilon = \eta/2$ for this k and all large n. This contradicts (6.9). Therefore μ is in E, and the proof is finished.

Corollary. *If a sequence (μ_n) in $\mathbf{P}$ converges to μ in $\mathbf{P}$ in* *the sense that (6.1) holds for all trigonometric polynomials ϕ, then (6.1) holds when ϕ is any bounded continuous function.*

Theorem 10 implies that the sequence converges weakly over $\mathbf{C_0(R)}$, and Theorem 11 that it converges over the space of all bounded continuous functions.

It might be hoped that Theorem 11 could be improved to say that the topology of $\mathbf{P_1}$ is the same as the weak* topology of $\mathbf{P}$ as part of the dual of the space of trigonometric polynomials. However, this is not true because the last topology is not defined by sequential convergence (Problem 2 below).

Problems

1. Prove the lemma used in the proof of Theorem 11.

2. Let E be the subset of $\mathbf{P}$ consisting of all unit point masses δ_t at points t such that $|t| \geq 1$. Show that E is closed in $\mathbf{P_1}$, but that δ_0 belongs to the closure of E for the weak* topology of $\mathbf{P}$ as part of the dual of trigonometric polynomials. [No *sequence* from E converges to δ_0. A trigonometric polynomial P has the property that for any positive ϵ there are arbitrarily large numbers t such that $|P(x+t) - P(x)| < \epsilon$ for all x.]

Chapter 3
Discrete and Compact Groups

1. Characters of discrete groups

This chapter is devoted to harmonic analysis on some compact abelian groups other than the circle group. We shall construct Haar measure for compact abelian groups, then prove the Pontryagin duality theorem for compact and discrete abelian groups, a theorem of Minkowski, Kolmogorov's extension theorem, and finally the Banach-Steinhaus theorem as a consequence of a theorem of Steinhaus about the set of distances between points in a set of positive measure.

Let $\mathbf{G}$ be any abelian group. A *character* of $\mathbf{G}$ is a homomorphism of $\mathbf{G}$ into $\mathbf{T}$. Among the characters is one that we denote by 1, with values $1(g) = 1$ for all g in $\mathbf{G}$. It is not obvious whether other characters exist in general.

A character is a function on $\mathbf{G}$. The characters of $\mathbf{G}$ themselves form a group under multiplication. The inverse of a character χ is $1/\chi = \bar{\chi}$, and the identity is the character 1. This group is also abelian, but it is natural to write it multiplicatively.

An element of $\mathbf{G}$ determines a character of its character group in a natural way: for g in $\mathbf{G}$ define $g(\chi) = \chi(g)$. Then $g(\chi\chi') = (\chi\chi')(g) = \chi(g)\chi'(g) = g(\chi)g(\chi')$, as required.

If $\mathbf{G}$ is an abelian Hausdorff topological group, its *continuous* characters form an abelian group $\hat{\mathbf{G}}$, called the *dual* of $\mathbf{G}$. Harmonic analysis treats groups that are locally compact; that is, having a neighborhood of 0 whose closure is compact. Then there is a natural way to define a locally compact topology in $\hat{\mathbf{G}}$. The duality theorem of Pontryagin states that *every continuous char-*

acter of $\hat{\mathbf{G}}$ *is obtained from an element of* $\mathbf{G}$ in the way just mentioned. A full proof of this theorem would take us too far from the subject of this book, but we shall prove it for groups that are discrete or compact.

Suppose that Γ is an abelian group with discrete topology. Our first result is that the characters of Γ (which are all continuous) separate points of Γ. Let x be any point of Γ different from the identity 0. If x has finite order k, define $\chi(x)$ to be a kth root of 1, different from 1, and $\chi(jx)$ to be $\chi(x)^j$ for all integers j. If x has infinite order, $\chi(x)$ can be any number of modulus 1 (but chosen different from 1). Then χ is a character of the subgroup Γ_0 of Γ generated by x. We shall extend χ to be a character of Γ.

Let $\mathfrak{F}$ be the family of all pairs (Γ_1, χ_1) where Γ_1 is a subgroup of Γ containing Γ_0, and χ_1 is a character of Γ_1 that agrees with χ on Γ_0. This family is partially ordered: $(\Gamma_2, \chi_2) >$ (Γ_1, χ_1) if Γ_2 contains Γ_1, and χ_2 extends χ_1. Any ordered subset of $\mathfrak{F}$ has an upper bound in $\mathfrak{F}$. By Zorn's lemma, there is a maximal element (Γ', χ') in $\mathfrak{F}$. If Γ' is not all of Γ, we can extend this pair by the argument just given. Choose y not in Γ'. If ky is in Γ' for some smallest positive integer k, let $\chi'(y)$ be any kth root of $\chi'(ky)$; if no positive multiple of y is in Γ', $\chi'(y)$ can be any number of modulus 1. Define $\chi'(x+jy) = \chi'(x)\chi'(y)^j$ for x in Γ' and all integers j. We verify that the definition is not ambiguous, and defines a character extending χ' to a group larger than Γ'. This contradicts the maximality of (Γ', χ'). Thus Γ' is Γ, χ' is a character on Γ, and $\chi'(x) \neq 1$.

It follows that characters separate any two points of Γ (Problem 1 below).

Let $\mathbf{K}$ be the dual of a discrete abelian group Γ. We have just shown that the elements of K separate points of Γ; it is trivial that elements of Γ, as functions on K, separate points of K. Give K the weakest topology such that all the elements λ of Γ, as functions on K, are continuous. A net (x_α) in K converges to x in this topology if and only if $x_\alpha(\lambda)$ converges to $x(\lambda)$ for every λ in Γ. This topology is compact. To prove this, construct a product of circles T_λ, one for each element λ of Γ. This space X is a compact Hausdorff space in the product topology (Tychonoff's theorem). Map K into X by setting $Tx = (x(\lambda))$. This mapping is one-one, and its range is closed (Problem 2 below). As a closed subset of a compact space, the range is compact. Convergence of (Tx_α) in X means convergence in each component, the same as the definition of convergence of (x_α) in K. Therefore K is homeomorphic to its image in X, so it is compact. The group K in this topology is $\hat{\Gamma}$.

The character of K defined by λ in Γ is continuous for this topology, by definition. It is to be proved that every continuous character of K arises from an element of Γ.

The *Baire field* of any compact Hausdorff space is the smallest σ-field that makes all continuous functions measurable. The Baire field is the same as the Borel field in any compact *metrizable* space, but in general it is smaller.

Let K be any compact abelian group (always assumed to have a Hausdorff topology). A *Haar measure* on K is a positive finite measure on the Baire field of K that takes positive values on non-empty open Baire sets, and is invariant under translations of the group. This measure is unique up to multiplication by positive constants. It can be normalized to have unit total mass.

Theorem 12. *Every compact abelian group K has a Haar measure.*

Let $\mathfrak{P}$ denote the set of probability measures on the Baire field of $\mathbf{K}$. This set is convex, and compact for the weak$*$ topology of measures. For μ in $\mathfrak{P}$ and x in $\mathbf{K}$, let μ^x be the measure in $\mathfrak{P}$ defined by

$$(1.1) \qquad \mu^x(E) = \mu(E + x)$$

for all Baire sets E. For every positive integer k,

$$(1.2) \qquad \nu_k = k^{-1} \sum_0^{k-1} \mu^{nx}$$

is in $\mathfrak{P}$. The set (ν_k) has an accumulation point ν for the weak$*$ topology of measures, and ν belongs to $\mathfrak{P}$. If $C(\mathbf{K})$, the space of continuous complex functions on $\mathbf{K}$, is separable, then ν is the limit of a subsequence (ν_{k_j}), so that

$$(1.3) \qquad \nu^x = \lim k_j^{-1} \sum_1^{k_j} \mu^{nx} = \lim (\nu_{k_j} + (\mu^{k_j x} - \mu)/k_j) = \nu.$$

That is, ν is invariant under translation by x. If ν is not the limit of any subsequence it is still true that given any finite subset $(\phi_1, \ldots, \phi_r)$ of $C(\mathbf{K})$, there is a subsequence (k_j) of integers such that

$$(1.4) \qquad \lim_j \int \phi_i \, d\nu_{k_j} = \int \phi_i \, d\nu$$

for each i. Given a continuous function ϕ, find such a subsequence so that (1.4) holds for $\phi(t)$ and its translate $\phi(t-x)$. The relation (1.3) holds then if we integrate the measures against $\phi(t)$, with the result that

(1.5) $$\int \phi \, d\nu^x = \int \phi \, d\nu.$$

Since this is true for any continuous ϕ, we have $\nu^x = \nu$.

Denote by $\mathfrak{P}_x$ the set of measures in $\mathfrak{P}$ that are invariant under translation by x. This set is non-empty, *-compact, and invariant under translation by all elements of **K**.

Let y be another element of **K**. Starting from any ν in $\mathfrak{P}_x$, we obtain in the same way a measure ρ that is invariant under translation by y, and still invariant under translation by x.

This construction can be repeated to show that every finite family of sets $\mathfrak{P}_{x_j}$ has non-empty intersection. Since $\mathfrak{P}$ is compact, the whole family has non-empty intersection. That is, there is a probability measure σ that is invariant under all translations of **K**. It is easy to see that non-empty open Baire sets have positive measure (Problem 3 below), so this is a Haar measure, and the theorem is proved.

Now let Γ be dual to **K**. The Fourier coefficients of a function f summable for σ on **K** are indexed by the elements of Γ, and are defined to be

(1.6) $$\hat{f}(\chi) = \int f \bar{\chi} \, d\sigma \qquad \text{(all } \chi \text{ in } \Gamma).$$

The Fourier series of f is

(1.7) $$f(x) \sim \sum_{\chi} \hat{f}(\chi) \, \chi(x).$$

A similar definition is made for Fourier-Stieltjes coefficients and series of measures on **K**.

A *trigonometric polynomial* on any abelian group is a finite linear combination of (continuous) characters. The *central coefficient* of a trigonometric polynomial is the coefficient of the character 1. Let $\mathcal{T}$ be the space of trigonometric polynomials on a compact group $\mathbf{K}$. $\mathcal{T}$ is an algebra, it contains constants, and is closed under complex conjugation. We assume now that the continuous characters of $\mathbf{K}$ separate points of $\mathbf{K}$. Then all the hypotheses of the Stone-Weierstrass theorem are satisfied; the conclusion is that $\mathcal{T}$ is dense in the space $\mathbf{C(K)}$, given the uniform topology.

Consequently the unicity theorem holds for measures on $\mathbf{K}$: if $\hat{\mu}(\chi) = 0$ for all characters χ of $\mathbf{K}$, then $\mu = 0$. For μ is the null functional on $\mathcal{T}$, and therefore on $\mathbf{C(K)}$, so μ is the null measure by Riesz' theorem.

Every character χ of $\mathbf{K}$ except the constant one has mean value 0 for the measure σ. For if y is any element of $\mathbf{K}$,

$$(1.8) \qquad \int \chi(x+y)\, d\sigma(x) = \chi(y) \int \chi(x)\, d\sigma(x).$$

Since σ is invariant under translation, the left side equals $\int \chi\, d\sigma$. If this integral is not 0, then $\chi(y) = 1$. This holds for all y; that is, $\chi = 1$. Thus the integral with respect to σ of a trigonometric polynomial is its central coefficient, which therefore can now be called its mean value.

It follows that the characters form an orthonormal system in $\mathbf{L^2(K)}$ (based on the measure σ). This system is complete, because the unicity theorem is true *a fortiori* in $\mathbf{L^2(K)}$ when it is known for measures on $\mathbf{K}$. Hence the Parseval relation holds:

$$(1.9) \qquad \int |f|^2\, d\sigma = \sum_{\chi} |\hat{f}(\chi)|^2$$

for all f in $\mathbf{L}^2(\mathbf{K})$.

All of this is true if we know that the continuous characters of $\mathbf{K}$ separate points of $\mathbf{K}$. Now we can show, under this hypothesis, that $\mathbf{K}$ is the dual of its character group Γ, given the discrete topology. Suppose that the group $\mathbf{K}'$ dual to Γ is larger than $\mathbf{K}$. The natural embedding of $\mathbf{K}$ in $\mathbf{K}'$ is continuous (Problem 4 below). Therefore the image is compact and closed. Trigonometric polynomials on $\mathbf{K}$, linear combinations of the characters from Γ, are dense in $\mathbf{C}(\mathbf{K})$; and on $\mathbf{K}'$, are dense in $\mathbf{C}(\mathbf{K}')$ (because they separate points of $\mathbf{K}'$). Find a function ϕ that is continuous and nonnull on $\mathbf{K}'$, but zero on $\mathbf{K}$; by squaring ϕ if necessary, we can assume that $\phi \geq 0$ everywhere. Let σ be the Haar measure of $\mathbf{K}$, and σ' that of $\mathbf{K}'$. Then

$$(1.10) \qquad \int \phi \, d\sigma = 0, \quad \int \phi \, d\sigma' > 0.$$

If (ϕ_n) is a sequence of trigonometric polynomials converging uniformly to ϕ on $\mathbf{K}'$, with central coefficients a_n, then the first relation implies that the a_n tend to 0, and the second that they have a positive limit. The contradiction shows that $\mathbf{K}'$ is not larger than $\mathbf{K}$. Thus we have shown that $\mathbf{K}$ is its own second dual, under the hypothesis that characters of $\mathbf{K}$ separate points.

Finally, let $\mathbf{K}$ be the dual of a discrete group Γ. Trivially, the characters of $\mathbf{K}$ that come from Γ separate points of $\mathbf{K}$ (as mentioned above). We want to show that $\mathbf{K}$ has no other characters. The trigonometric polynomials containing only characters from Γ form an algebra that is dense in $\mathbf{C}(\mathbf{K})$. If $\mathbf{K}$ had some other character χ, then χ would be orthogonal in $\mathbf{L}^2(\mathbf{K})$ to the characters from Γ, but uniformly approximable by linear combinations of them. This is impossible. Therefore Γ is its own second dual.

We know now that every discrete abelian group is its own second dual; and the same is proved for compact abelian groups having the property that continuous characters separate points. This is always true, as we establish in the next section.

Problems

1. Show that if characters of Γ separate 0 from every other element, then characters separate every pair of distinct elements.

2. Prove the facts needed above: the mapping T from **K** into **X** is one-one, and its range is closed. [If $(x_\alpha(\lambda))$ converges in **X** to a limit $(x(\lambda))$, then x (a function on Γ) is a character of Γ and x_α tends to x in **K**.]

3. The measure σ on **K** was asserted to be a Haar measure. Show that it is positive on every non-empty open Baire set. [Let E be a non-empty open set. Finitely many translates of E cover **K**.]

4. Supply the point missing above: **K** is continuously embedded in **K′**.

5. Write down a careful proof of (1.5) above.

6. Find a second proof of the existence of Haar measure on a group **K** about which it is known that continuous characters separate points. For trigonometric polynomials P, define $F(P)$ to be the central coefficient of P. Show that F is a linear functional with bound 1, which therefore has an extension to all of **C(K)** with the same bound. Apply the theorem of F. Riesz. [Suppose there is a trigonometric polynomial P such that $||P||_\infty < 1$ but $F(P) = 1$. Let λ be a nonzero character appearing in P, and y an

element of K such that $\lambda(y) \neq 1$. Average the functions $P(x + ny)$ to obtain a new trigonometric polynomial like P, but not containing λ.]

2. Characters of compact groups

Theorem 13. *Continuous characters of any compact abelian group* K *separate points of* K.

Finding characters of a discrete group involved only showing that a character on any subgroup could be extended to a slightly larger subgroup. Then Zorn's lemma completed the proof. In other words, we only had to establish that there is no algebraic impediment to extension. But in the case of a compact group, the character is supposed to be continuous. Given x in K it is not easy to find a non-trivial continuous character even on the smallest closed subgroup containing x. If χ is such a function, then $\chi(n_j x)$ must tend to 1 for every sequence (n_j) such that $n_j x$ tends to 0 in K. This is a Diophantine condition that is difficult to verify because we know nothing about the structure of K. Indeed the result can only be true because compact abelian groups have simpler structure than we have any reason to hope.

The proof will use operator theory in the Hilbert space $\mathbf{L}^2(\mathbf{K})$, based on Haar measure σ.

For any y in the group, let T_y be the translation operator: $T_y f(x) = f(x + y)$. Then T_y is a unitary operator in $\mathbf{L}^2(\mathbf{K})$. For any f in $\mathbf{L}^2(\mathbf{K})$, $T_y f$ depends continuously on y. (The proof is like that on the circle, since trigonometric polynomials are dense in $\mathbf{L}^2(\mathbf{K})$.)

Let h be any element of $\mathbf{L}^2(\mathbf{K})$. Define $S_h f = h * f$. Then S_h is a continuous linear operator in $\mathbf{L}^2(\mathbf{K})$. This convolution operator commutes with the translation operators T_y.

Lemma. S_h *is normal and compact.*

Let $\tilde{h}$ be the function $\overline{h}(-x)$. It is easy to verify that the adjoint of S_h is $S_{\tilde{h}}$. Since two convolutions commute, S_h is normal.

To show that S_h is compact it suffices to prove that if (f_n) is a sequence in $\mathbf{L}^2(\mathbf{K})$ that converges weakly to f, then $S_h f_n$ converges in norm to $S_h f$. We have

$$(2.1) \qquad ||S_h f - S_h f_n||^2 = \int |h*(f-f_n)|^2 \, d\sigma.$$

Since h is in $\mathbf{L}^2(\mathbf{K})$, weak convergence of f_n to f implies that the integrand tends to 0 pointwise. A weakly convergent sequence is bounded; therefore it follows from the Schwarz inequality that the integrand is uniformly bounded in n. Thus the integral tends to 0.

Now we appeal to the theory of compact operators. The spectrum of such an operator is a sequence of nonzero complex numbers (λ_j) tending to 0, together with the point 0 (unless there are only finitely many λ_j). Each λ_j is an eigenvalue, and the corresponding eigenspace is finite-dimensional. Finally, 0 may or may not be an eigenvalue, and if it is, its eigenspace may be infinite-dimensional.

If S_h has only 0 in its spectrum, then the operator is null and h is the null function. (This is a property of normal operators, obvious from the spectral theorem, if not proved in an elementary way.) Otherwise let λ be a nonzero eigenvalue and let E be the eigenspace corresponding to λ: that is, all f in $\mathbf{L}^2(\mathbf{K})$ such that $h*f = \lambda f$. This is a closed subspace, invariant under translations, and (this is the essential point of the proof) finite-dimensional. Because E is finite-dimensional, it has a subspace E' different from (0) that is invariant under translations, and is minimal with this property.

Each translation operator T_x acts as a unitary operator in E'. Furthermore, T_x in E' has only one eigenvalue $\tau(x)$; for if it had more than one, its eigenspace in E' would be a proper subspace of E' that is invariant under translation. It is easy to verify that $\tau(x)$ is a character on $\mathbf{K}$, and is continuous. The only thing left to prove is that given x not the identity of $\mathbf{K}$, we can choose h in such a way that $\tau(x) \neq 1$.

Find k in $\mathbf{L}^2(\mathbf{K})$ that is not invariant under T_x. (The characteristic function of a sufficiently small neighborhood of 0 will do.) Then take $h = k - T_x k$. If λ is any nonzero eigenvalue of S_h, and f an eigenfunction, we have

$$(2.2) \qquad h*f = k*f - (T_x k)*f = \lambda f.$$

Now $(T_x k)*f = k*(T_x f)$. If $T_x f = f$, then the middle expression in (2.2) vanishes, whereas the right side is not the null function. Hence no nonzero element of the eigenspace E corresponding to λ is invariant under T_x. This means that the eigenvalue of T_x acting in a minimal invariant subspace of E is not 1, and the proof is finished.

Problems

1. Show that normalized Haar measure σ on a compact abelian group $\mathbf{K}$ is unique. [Calculate its Fourier-Stieltjes coefficients.]

2. Show that every closed subspace of $\mathbf{L}^2(\mathbf{K})$ invariant under translations is the closed span of the characters it contains.

(This is a version of Wiener's theorem proved on the circle group in Chapter 1.) Deduce that every minimal closed invariant subspace of $L^2(K)$ is one-dimensional. (Thus the minimal subspace E' in the proof above was actually one-dimensional.)

3. Let K be a compact abelian group with dual Γ and K_0 a closed subgroup. Show that the dual of K_0 is the quotient group Γ/Γ_0, where Γ_0 is the annihilator of K_0. Formulate and prove an analogous statement about subgroups of Γ.

4. Let χ be a continuous character of a closed subgroup of a compact abelian group K. Show that χ can be extended to be a continuous character of K.

3. Bochner's theorem

Theorem 14. *Let Γ and K be dual abelian groups, respectively discrete and compact. Every positive definite function on Γ is the Fourier-Stieltjes transform of a positive measure on K. Every continuous positive definite function on K has absolutely convergent Fourier series with non-negative coefficients.*

Let ρ be a positive definite function on Γ. That is, if c is any complex function supported on a finite subset of Γ, then

$$(3.1) \qquad \sum_{\lambda, \tau} \rho(\lambda - \tau) c_\lambda \bar{c}_\tau = \sum_\lambda \rho(\lambda) c * \tilde{c}(\lambda) \geq 0,$$

where $\tilde{c}(\lambda) = \bar{c}(-\lambda)$ as always. We may assume that $\rho(0) = 1$. The proof of Herglotz' theorem in Chapter 1, and of Bochner's theorem in the last chapter, relied on the Fejér kernel. In the present more general context we have to construct an approximate identity with the necessary properties.

A positive definite function ρ is bounded (with bound

$\rho(0)$). It follows that the sum in (3.1) converges absolutely, and the inequality still holds, if c is any function summable on Γ.

Let V be a neighborhood of the identity in $\mathbf{K}$, and e its characteristic function. Then $e * \tilde{e}$ is a continuous function whose Fourier sequence is $|\hat{e}|^2$. Therefore the Fourier series of $e * \tilde{e}$ converges absolutely, and its Fourier coefficients $c_\lambda = |\hat{e}(\lambda)|^2$ are non-negative. Finally define $k = \kappa |e * \tilde{e}|^2$, where κ is the positive number such that $||k||_1 = 1$. Actually $e * \tilde{e}$ is non-negative, so the absolute value is redundant; but we emphasize the fact that $\hat{k} = \kappa c * \tilde{c}$. Note that k vanishes outside $V - V$ (the set of all $x - y$ where x and y belong to V).

In (3.1) replace c_λ by $c_\lambda \lambda(x)$, where c_λ is the Fourier coefficient just introduced, and x is in $\mathbf{K}$. The series converges absolutely and we have

$$(3.2) \quad h(x) = \kappa \sum_\lambda \rho(\lambda) c * \tilde{c}(\lambda) \lambda(x) = \sum_\lambda \rho(\lambda) \hat{k}(\lambda) \lambda(x) \geq 0.$$

This function $h = h_V$ on $\mathbf{K}$ has norm 1 in $\mathbf{L}^1(\mathbf{K})$ for each V. The neighborhoods V of 0 in $\mathbf{K}$ form a system directed by inclusion, and so $(h_V d\sigma)$ is a net in the set of probability measures on $\mathbf{K}$.

By compactness of the set of probability measures, the net has an accumulation point μ. Therefore, for each λ in Γ, $\hat{h}_V(\lambda)$ is as close as we please to $\hat{\mu}(\lambda)$ for all neighborhoods V of 0 that are sufficiently small (that is, contained in a corresponding neighborhood V_1 of 0).

Fix λ in Γ, and let δ be a positive number. The set of all x such that $|\lambda(x) - 1| < \delta$ is a neighborhood W of 0 in $\mathbf{K}$. There is a neighborhood V_1 of 0 such that $V_1 - V_1$ is contained in W (this is an elementary fact about topological groups). Let V be *any* neighborhood of 0 contained in V_1. Since k_V vanishes outside W, where λ is close to 1,

$$(3.3) \qquad 1 \geq \hat{k}_V(\lambda) = \int k \bar{\lambda}_V \, d\sigma \geq 1 - \delta.$$

Hence $\hat{h}_V(\lambda) = \rho(\lambda)\,\hat{k}_V(\lambda)$ is within δ of $\rho(\lambda)$ for all V contained in V_1. Thus $\rho(\lambda)$ is the only accumulation point of $(\hat{h}_V(\lambda))$, so that $\hat{\mu}(\lambda) = \rho(\lambda)$, and this proves Bochner's theorem for Γ.

Now let f be a continuous positive definite function on $\mathbf{K}$, with Fourier coefficients a_λ. Part of the proof given in Chapter 2, Section 5, can be repeated, with some simplification. For any complex numbers c_j and elements x_j of $\mathbf{K}$, M in number, we have

$$(3.4) \qquad \sum f(x_j - x_k)\, c_j \bar{c}_k \geq 0.$$

Let g be any function in $\mathbf{L}^2(\mathbf{K})$ with norm 1, and set $c_j = g(x_j)$ for each j:

$$(3.5) \qquad \sum f(x_j - x_k)\, g(x_j)\, \bar{g}(x_k) \geq 0.$$

Integrate this inequality over $\mathbf{K}$ with respect to each variable $d\sigma(x_j)$ $(j = 1, \ldots, M)$:

$$(3.6) \quad Mf(0) + M(M-1) \int f(x - y)\, g(x)\, \bar{g}(y)\, d\sigma(x)\, d\sigma(y) \geq 0.$$

Since M is arbitrary, the integral itself must be non-negative; that is, f is positive definite in an integral sense. Taking for g a character shows that $\hat{f}(\lambda) \geq 0$ for all λ in Γ.

Any continuous function with non-negative Fourier coefficients must have absolutely convergent Fourier series (Problem 2 below). This completes the proof.

Problems

1. Using Bochner's theorem, show that the product and the convolution of two continuous positive definite functions on **K** are positive definite.

2. Show that a continuous function f on **K** with non-negative Fourier coefficients has absolutely convergent Fourier series. [It suffices to know that f can be approximated uniformly by trigonometric polynomials whose coefficients are non-negative. The Fejér kernel accomplishes this on the circle group; on **K**, the approximate identity constructed in the first part of the proof will do the same.]

4. Examples

(a) Let Γ be the group **Z** of integers. A character χ of **Z** is determined by its value $\chi(1)$, a number $\exp it$ of modulus 1, for then $\chi(n) = \exp nit$ for every integer n. Different numbers $\exp it$ determine different characters, so this is a correspondence between the points of **T** and the characters of **Z**. The multiplication in **T** corresponds to multiplication of characters. Since **Z** is a discrete space, every character is continuous. Thus $\hat{\mathbf{Z}}$ is identified with **T**. Finally, convergence in **T** is the same as pointwise convergence over **Z**, so **T**, in its ordinary topology, is the compact group dual to **Z**.

It follows from the duality theorem that the characters of **T** are all obtained from **Z**: $\chi_n(e^{it}) = e^{nit}$ for n in **Z**. This is not quite obvious.

(b) Take for Γ the set $(0, 1, \ldots, q-1)$, where q is an integer greater than 1. Define addition in Γ by reduction modulo q. This is a finite group $\mathbf{Z}_q$, isomorphic to the quotient group $\mathbf{Z}/q\mathbf{Z}$. The group is cyclic, with generator 1. For any character χ, $\chi(1)$ must be a qth root of unity: $\exp 2\pi i r/q$ where r has one of the values $0, \ldots, q-1$. Therefore χ has the form

$$(4.1) \qquad \chi_r(k) = \exp 2\pi k i r/q.$$

Obviously multiplying characters has the effect of adding their indices r modulo q. Thus the mapping from r to χ_r is an isomorphism of $\mathbf{Z}_q$ with its dual group.

(c) Knowing the character group of $\mathbf{Z}$ enables us easily to find that of $\mathbf{Z}^2$, the group of pairs (m, n) of integers with addition in each component as the group operation. Given a character χ of $\mathbf{Z}^2$, define a character χ_1 of $\mathbf{Z}$ by setting $\chi_1(m) = \chi(m, 0)$. χ_1 must have values $\exp mit$ for some real t. Similarly, $\chi_2(n) = \chi(0, n)$ must equal $\exp niu$ for some u. Since χ is a homomorphism, $\chi(m, n) = \exp i(mt + nu)$ for all integer pairs. Thus the characters of $\mathbf{Z}^2$ are in correspondence with points (e^{it}, e^{iu}) of $\mathbf{T}^2$, and this is an isomorphism of $\hat{\mathbf{Z}}^2$ with $\mathbf{T}^2$. Therefore the dual of $\mathbf{T}^2$ is naturally identified with $\mathbf{Z}^2$: $\chi(e^{it}, e^{iu}) = \exp i(mt + nu)$ for some integers m, n.

This procedure can be carried on to show that $\mathbf{Z}^n$ and $\mathbf{T}^n$ are dual to each other for each positive integer n, and more generally, to show that the group dual to a product of groups is the product of their dual groups. We shall not state this result formally because the notation is complicated, and in any particular case it is obvious.

(d) We shall be interested in the *infinite sum* of groups Γ_j, $j = 1, 2, \ldots$. Let Γ be the set of all sequences $k = (k_1, k_2, \ldots)$ where each k_j belongs to Γ_j, and where *all k_j from some point on are* 0. The group operation is addition in each component. Give Γ the discrete topology. We shall describe its dual.

The argument of (c) applies here. For χ any character of Γ, $\chi(k_1, 0, 0, \ldots)$ determines a character χ_1 of Γ_1, which is an element of the group K_1 dual to Γ_1. In the same way we find characters χ_j of Γ_j for $j = 2, 3, \ldots$ so that

(4.2) $$\chi(k) = \prod \chi_j(k_j).$$

For any k, $k_j = 0$ and $\chi_j(k_j) = 1$ for all j from some point on; therefore the product consists of factors that are 1 except for finitely many indices. Conversely, for any sequence $\chi = (\chi_1, \chi_2, \ldots)$, where each χ_j is a character of Γ_j, (4.2) defines a character of Γ. Thus the dual of Γ is realized by the set K of such sequences.

The group operation in K is multiplication in each component. The topology of K as the dual of Γ is the same as pointwise convergence in each component, which is the ordinary product topology. K is called the *infinite product* of the groups K_j.

A group of particular interest is the infinite sum of groups Z_2, whose elements are sequences $k = (k_1, k_2, \ldots)$ where each k_j is 0 or 1, and equal to 1 for only finitely many indices; addition is performed modulo 2 in each component. The dual group Z_2^∞ is the infinite product of groups Z_2, whose elements are sequences $x = (x_1, x_2, \ldots)$ where each x_j is 0 or 1, with no restriction; and addition is again modulo 2 in each component. The pairing is given by the formula $x(k) = \exp \pi i \sum k_j x_j$. Every character takes the values 1 and -1 only.

Problems

1. The *symmetric difference* of two sets is the set of points that belong to one but not both the sets. Show that the family of all subsets of a set S is a group under this operation. Show that the family of all *finite* subsets of S is also a group. Take S to be the set of positive integers. Show how to identity the first family with Z_2^∞, and the second family with the infinite sum of groups Z_2.

2. Write down explicitly the definitions of Fourier coefficient and Fourier series for functions on T^2. Write down the Parseval relation for this group.

3. Take for Γ the discrete additive group of dyadic rational numbers, that is, all r/s where r and s are integers and s is a power of 2. Describe the dual of this group.

4. Show that Z_2^∞ is metrizable. That is, a metric can be defined in it giving rise to its topology. Show that every open set is a Baire set, so that the Borel and Baire fields coincide.

5. The group Z_2^∞ can be mapped to the unit interval by setting

$$y = \sum_1^\infty x_n/2^n.$$

If the countable set of x that end in 1's is removed, the mapping is one-one onto $[0, 1)$. Show that the Borel field of Z_2^∞ is carried onto the Borel field of the interval. Show that σ is carried to Lebesgue measure on the interval. Describe characters of Z_2^∞ as functions on the interval.

5. Minkowski's theorem

This section reproduces the beautiful proof by C. L. Siegel of a theorem of Minkowski that is important in the theory of numbers.

For positive integers n, $\mathbf{T}^n$ is the n-dimensional torus, the product of n copies of the circle $\mathbf{T}$. We write its elements as $X = (\exp 2\pi i x_1, \ldots, \exp 2\pi i x_n)$, where the x_j are real numbers. This is a compact abelian group when elements are multiplied in each component. Its normalized Haar measure σ_n is the n-fold product of σ on $\mathbf{T}$. This group is the dual of $\mathbf{Z}^n$; the duality is given by

$$(5.1) \qquad \exp 2\pi i K \cdot X = \exp 2\pi i \sum k_j x_j$$

where $K = (k_1, \ldots, k_n)$ is in $\mathbf{Z}^n$.

Minkowski's theorem. *Let $\mathbf{C}$ be a convex body in $\mathbf{R}^n$ that is symmetric about the origin. If its volume V is greater than 2^n, then $\mathbf{C}$ contains a lattice point other than the origin.*

(We shall not prove that $\mathbf{C}$ is a measurable set; this fact is usually obvious in applications.)

If $\mathbf{C}$ is the cube consisting of all x in $\mathbf{R}^n$ such that $|x_j| < 1$ for each j, then $\mathbf{C}$ contains no lattice point except the origin, and V is exactly 2^n. Thus the constant cannot be improved.

Assume that $\mathbf{C}$ is bounded. If the theorem holds under this hypothesis, then it holds in general. Denote the characteristic function of $\mathbf{C}$ by ϕ. Then we define

$$(5.2) \qquad f(X) = \sum_K \phi(2X - 2K)$$

where K ranges over the lattice points of $\mathbf{R}^n$. Since $\mathbf{C}$ is bounded, the sum has only finitely many nonzero terms for each X. The function f is periodic with period 1 in each component, and is bounded. Let E be the cube consisting of all X with $0 < x_j < 1$ for each j. The Fourier coefficients of f are

$$\hat{f}(K) = \int_E f(X) \exp{-2\pi i K \cdot X}\, d\sigma_n(X)$$

(5.3)
$$= \int_{\mathbf{R}^n} \phi(2X) \exp{-2\pi i K \cdot X}\, d\sigma_n(X)$$

$$= 2^{-n} \int \phi(X) \exp{-\pi i K \cdot X}\, d\sigma_n(X).$$

The Parseval relation, written for functions with period 1, gives

(5.4)
$$\sum |\hat{f}(K)|^2 = \int_E |f|^2\, d\sigma_n(X).$$

The right side equals

(5.5) $\displaystyle\int_E f(X) \sum \phi(2X - 2K)\, d\sigma_n(X) = \int_{\mathbf{R}^n} f(X)\, \phi(2X)\, d\sigma_n(X)$

because f is periodic. Setting in the definition of f again gives

(5.6)
$$\sum \int_{\mathbf{R}^n} \phi(2X - 2K)\, \phi(2X)\, d\sigma_n(X) =$$

$$2^{-n} \sum \int_{\mathbf{R}^n} \phi(X - 2K)\, \phi(X)\, d\sigma_n(X) = 2^{-n} \sum \int_{\mathbf{C}} \phi(X - 2K)\, d\sigma_n(X).$$

If X and $X - 2K$ are both in $\mathbf{C}$, then $2K - X$ is in $\mathbf{C}$ ($\mathbf{C}$ is symmetric), and $(X + 2K - X)/2 = K$ is in $\mathbf{C}$ ($\mathbf{C}$ is convex). Assume that $\mathbf{C}$ contains no lattice point except the origin. Then the sum on the right side of (5.6) must have only one term different from 0, the one with $K = 0$. That term is $2^{-n} V$. After canceling a factor 2^{-n}, (5.3), (5.4) and (5.6) lead to

$$(5.7) \qquad 2^{-n} \sum \left| \int \phi(X) \exp -\pi i K \cdot X \, d\sigma_n(X) \right|^2 = V.$$

The term on the left with $K = 0$ is $2^{-n} V^2$, so that $2^{-n} V^2 \leq V$. This proves the theorem.

The equality (5.7) gives more information than the inequality of the theorem; Siegel uses it to obtain finer results about lattice points in particular convex bodies.

Problems

1. If **C** is the body of Minkowski's theorem and $V = 2^n$, then there is a lattice point other than the origin in **C** or on its boundary.

2. Use Minkowski's theorem to prove this fact: if $k > 0$ and a, b, c, d are real numbers with $|ad - bc| \leq 1$, then there are integers m, n not both 0 such that $|am + bn| \leq k$ and $|cm + dn| \leq k^{-1}$. Deduce that for every real number a there are infinitely many integer pairs m, n such that

$$\left| a + \frac{n}{m} \right| \leq \frac{1}{m^2}.$$

[This result can be proved directly, and it can be improved, but it illustrates the application of Minkowski's theorem to Diophantine problems.]

3. Let **C** be a closed symmetric convex subset of $\mathbf{R}^n$ that is unbounded. Show that **C** contains a subspace.

6. Measures on infinite product spaces

This section proves the *extension theorem* of Kolmogorov, using Bochner's theorem instead of conventional measure theory.

For each positive integer n let $\mathbf{K_n}$ be a compact abelian group. Call their infinite product $\mathbf{K}$. For each positive integer k let S_k denote the family of Baire sets in $\mathbf{K}$ that depend only on the first k components; that is, E belongs to S_k if whenever $x = (x_1, x_2, \ldots)$ belongs to E, so do all points $y = (y_1, y_2, \ldots)$ such that $x_j = y_j$ for $1 \le j \le k$. Each S_k is a σ-field.

Theorem 15. *Let μ be a function defined on the union of the σ-fields S_k such that μ is a probability measure on each S_k. Then μ has a unique extension to a probability measure on the Baire field of* $\mathbf{K}$.

For each n let $\mathbf{\Gamma_n}$ be the discrete group dual to $\mathbf{K_n}$, and let $\mathbf{\Gamma}$ be their infinite sum, which is dual to $\mathbf{K}$. A character χ of $\mathbf{K}$ is a function depending on only finitely many components x_j, which is therefore measurable with respect to some S_k. The integral

$$(6.1) \qquad \rho(\chi) = \int \bar{\chi} \, d\mu$$

is defined, and does not depend on which field S_k is used to calculate it, provided that k is large enough to make χ measurable. It is easy to verify that ρ is positive definite on $\mathbf{\Gamma}$, because the defining relation (formula (3.1)) involves only finitely many characters of $\mathbf{K}$, each depending on only finitely many variables. By Bochner's theorem, ρ is the Fourier-Stieltjes sequence of a probability measure ν on Baire sets of $\mathbf{K}$. The unicity theorem for measures on a finite product of groups $\mathbf{K_n}$ shows that the restriction of ν to each S_k is the same as the restriction of μ to S_k, so that ν is an extension of μ. This proves the theorem.

As a special case of the theorem, suppose that each $\mathbf{K_n}$ carries a probability measure μ_n. On each finite product $\mathbf{K_1} \times \ldots \times \mathbf{K_n}$ suppose that a product measure $\mu_1 \times \ldots \times \mu_n$ has been defined, and the Fubini theorem proved. Then Theorem 15 yields a measure on the infinite product space $\mathbf{K}$ having the expected relation to the measures μ_n.

These are results that apparently have nothing to do with harmonic analysis. It is interesting that the special result just stated about product measures is indeed a theorem in measure theory that requires no hypotheses on the measure spaces that enter into the product; but Theorem 15 is not true in such generality, and some topological hypothesis is needed. The theorem remains true, of course, if each $\mathbf{K_n}$ is a measurable space that is isomorphic to the Baire structure of some compact abelian group. For example, each $\mathbf{K_n}$ can be a copy of the real line.

For convenience Theorem 15 was formulated for countable infinite products of compact groups. It is easy to extend the definitions and proof to arbitrary products.

7. Continuity of seminorms

This section uses integration over a compact abelian group to prove classical results of functional analysis.

A *seminorm* on a vector space is a real function N on the space such that $N(x) \geq 0$, $N(tx) = |t| N(x)$, and $N(x + y) \leq N(x) + N(y)$, for all x, y in the space and all scalars t. This would be a norm if in addition $N(x) \neq 0$ for all $x \neq 0$.

Theorem 16. *A Borel seminorm on a real Banach space is continuous.*

This theorem is due to Banach, but is not well known. The usual, less general version states that a lower semicontinuous seminorm on a Banach space is continuous, and the proof is based on Baire's theorem. Our proof will use, instead, the

Theorem of Steinhaus. *If a set E in a compact abelian group* **K** *is measurable and has positive Haar measure, then there is a neighborhood V of the identity in* **K** *such that every element of V is $x - y$ for some x and y in E.*

Steinhaus stated his result in Euclidean space, where it says that every sufficiently small positive number is the distance between some two points of E. The following proof is simpler than the original one.

Let f be the characteristic function of E. Then the integral

$$(7.1) \qquad F(z) = \int f(z + y) f(y) \, d\sigma(y)$$

is a continuous function (because translation in $L^2(K)$ is continuous), with $F(0) = \sigma(E) > 0$. Hence F is positive on a neighborhood V of 0. If z is in V, the integrand is not everywhere 0; hence for some y, both y and $x = z + y$ belong to E. Then $z = x - y$, as was to be proved.

For positive integers n, denote by V_n the set of all $x = (x_1, x_2, \ldots)$ in $\mathbf{Z}_2^\infty$ such that $x_j = 0$ for $j = 1, \ldots, n$. This is a cyclinder set, a neighborhood of 0, and the family (V_n) is a base for neighborhoods of 0. (Every open set that contains 0 contains some V_n.) If δ_k denotes the element of $\mathbf{Z}_2^\infty$ with $\delta_k(j) = 1$ for $j = k$, $= 0$ for all $j \neq k$, then V_n contains δ_k for all $k > n$.

Now let N be a Borel seminorm on a Banach space **X**. (N is measurable for the smallest σ-field containing all open sets of

X.) If N is not continuous, there are elements w_j of **X** such that $\|w_j\| = 1/j^2$ but $N(w_j) \geq j$. For any sequence x of 0's and 1's define

$$(7.2) \qquad\qquad h(x) = \sum_1^\infty x_j w_j,$$

a sum convergent in norm in **X**. h is a function from $\mathbf{Z}_2^\infty$ into **X**. For each positive integer k,

$$(7.3) \qquad\qquad h_k(x) = \sum_1^k x_j w_j$$

is a continuous function of x, because each component x_j depends continuously on x. Since h is the uniform limit of (h_k), h is continuous. Therefore the composition $N(h(x))$ is a Borel mapping from $\mathbf{Z}_2^\infty$ to real numbers.

Let E_K be the set of x where $N(h(x)) \leq K$, a Borel set in $\mathbf{Z}_2^\infty$. The union of the E_K is all of $\mathbf{Z}_2^\infty$; therefore some E_K has positive measure. We fix K and call this set E. For all j we have

$$(7.4) \qquad\qquad j \leq N(w_j) = N(h(\delta_j)).$$

By the theorem of Steinhaus, the difference set $E - E$ contains one of the sets V_n, and therefore contains δ_j for all large enough j. Thus for every large j we can write $\delta_j = r - s$ for some r, s in E. These elements r and s of $\mathbf{Z}_2^\infty$ have the same components except at the jth place, where they differ. Since $r - s = s - r$, we may suppose that $r_j = 1$ and $s_j = 0$. Then from (7.2), $h(r - s) = h(r) - h(s)$. Since r and s belong to E, (7.4) can be continued:

$$(7.5) \qquad N(h(\delta_j)) = N(h(r) - h(s)) \leq N(h(r)) + N(h(s)) \leq 2K.$$

The left side exceeds j but the right side is fixed. The contradiction shows that N is continuous.

As a corollary we obtain the

Banach-Steinhaus Theorem. *Let* $T_1, T_2, \ldots$ *be continuous linear operators from a Banach space* **X** *into a normed vector space* **Y**. *If* $(T_n x)$ *is a bounded sequence in* **Y** *for each* x *in* **X**, *then the norms* $\|T_n\|$ *all lie under a common bound.*

Define $N(x)$ to be the supremum of the non-negative numbers $\|T_n x\|$, finite for each x. For each n, this norm is a continuous function of x, and therefore N is a Borel function. It is easy to check that N is a seminorm. By Theorem 16, N is continuous, and this is equivalent to the statement of the theorem.

Problems

1. Prove this result, analogous to Theorem 16. Let N be a non-negative Borel function on a real Banach space **X** such that $N(tx) = N(x)$ and $N(x - y) \leq \max(N(x), N(y))$ for x, y in **X** and $t > 0$. Show that N is bounded. Show that N assumes its upper bound A, by applying the first result to $(A - N)^{-1}$.

2. Use the result of Problem 1 to show that if the Banach space **X** is the union of an increasing sequence of Borel subspaces, then one of the subspaces is equal to **X**.

3. Use the result of Problem 1 to prove this result. Let $A_1, A_2, \ldots$ be linear operators in a Banach space **X**. For each x in **X** and F in the dual space, suppose that $\sum F(A_n x) z^{-n}$ converges for $z \neq 0$, and the sum has only a pole at 0. (The order of the pole depends on x and F.) Then actually $A_n = 0$ for all but finitely many n.

Chapter 4
Hardy Spaces

1. $H^p(T)$

For $1 \leq p \leq \infty$, $H^p(T)$ is the subspace of $L^p(T)$ consisting of f such that $a_n(f) = 0$ for all $n < 0$. This subspace is closed in $L^p(T)$, and *-closed if $p > 1$ (when $L^p(T)$ is a dual space). The functions of $H^p(T)$ have Fourier series

$$(1.1) \qquad f(e^{ix}) \sim \sum_0^\infty a_n e^{nix}.$$

Thus the harmonic extension

$$(1.2) \qquad F(z) = \sum_0^\infty a_n z^n \qquad (z = re^{ix})$$

is actually analytic.

In Chapter 1 we showed that F is obtained from f by means of the Poisson integral, and we reconstructed the boundary function f from F. If we set $f_r(e^{ix}) = F(re^{ix})$, then f_r converges to f in norm as r increases to 1 for finite p; if $p = \infty$, the convergence is in the *-topology of $L^\infty(T)$ as the dual of $L^1(T)$. For continuous f, the convergence is uniform. If f is replaced by a measure, then f_r tends to the measure in the *-topology of measures.

The next step was to start from a function F, now given as a power series (1.2), and find f under the hypothesis that

$$(1.3) \qquad \int |F(re^{ix})|^p \, d\sigma(x) = \|f_r\|_p^p \leq K < \infty \qquad (0 < r < 1).$$

For $1 < p \leq \infty$ we showed that there is an f in $L^p(T)$ whose

Poisson integral is F. If $p = 1$, the boundary object might be a measure.

We are interested now in results about *analytic* functions that are not true in general for harmonic functions.

A first, remarkable fact is that the exceptional situation of the exponent $p = 1$ disappears for analytic functions. This is the

Theorem of F. and M. Riesz. *If μ is a measure of analytic type, that is if $a_n(\mu) = 0$ for all $n < 0$, then μ is absolutely continuous. Equivalently, if F is analytic in the unit disk and (1.3) holds with $p = 1$, then F is the Poisson integral of a function in $\mathbf{H}^1(\mathbf{T})$.*

A proof will be given in the next section.

A pervasive and fundamental phenomenon is that the values of an analytic function cannot vanish on a large set unless the function is null. The first version of this statement is the result that the zeros of a nonnull analytic function cannot accumulate within the region of analyticity. Of course the zeros can accumulate on the boundary. We shall see now that the boundary function of a nonnull function in any $\mathbf{H}^p(\mathbf{T})$ cannot vanish on a set of positive measure. In different degrees of generality and precision, this result is due to Fatou, A. Ostrowski, F. and M. Riesz, and Szegö. Here is a theorem of this type with a proof in Hilbert space.

Theorem 17. *If f is in $\mathbf{H}^2(\mathbf{T})$ and vanishes on a set of positive measure, then f is null.*

The statement, and the proof to follow, mention only Fourier coefficients. This is properly a theorem of harmonic analysis rather than of function theory.

If f is not null, then after division by a power of χ we may suppose that $a_0(f) \neq 0$, and indeed $= 1$. Let $\mathfrak{F}$ be the closure of the subset of $\mathbf{H^2(T)}$ consisting of all functions

$$(1.4) \qquad\qquad (1 + b_1\chi + \ldots + b_n\chi^n)f$$

where the b_j are complex numbers and n varies. Each function in $\mathfrak{F}$ has mean value 1, because that is true of each product (1.4). Moreover if f vanishes on a set E, the same is true of the functions (1.4), and therefore (modulo sets of measure 0) of all the functions of $\mathfrak{F}$. We assume that E has positive measure and derive a contradiction.

$\qquad\mathfrak{F}$ is a closed convex subset of the Hilbert space $\mathbf{L^2(T)}$, and so has an element g of smallest norm. It is easy to see that $g + t\chi^n g$ is in $\mathfrak{F}$ for any complex number t and positive integer n. Hence the expression

$$(1.5) \quad \int |g + t\chi^n g|^2 \, d\sigma \ = \ \int |g|^2 \, d\sigma + 2\Re\left[t\int |g|^2 \chi^n \, d\sigma \right] + |t|^2 \int |g|^2 \, d\sigma$$

has a minimum at $t = 0$. This is a quadratic expression in t; by elementary means it follows that

$$(1.6) \qquad\qquad \int |g|^2 \chi^n \, d\sigma \ = \ 0.$$

This is proved for n any positive integer; taking the complex conjugate of (1.6) shows that it holds also for negative n. That is, all the Fourier coefficients of $|g|^2$ (a summable function) are 0 except the mean value. Hence g is constant. Since g was assumed to vanish on the set E of positive measure, $g = 0$. But g is in $\mathfrak{F}$ and has mean value 1. This contradiction proves the theorem.

$\qquad$In their paper of 1916, F. and M. Riesz proved a function-theoretic statement that combines the two theorems mentioned in this section. Let f be an analytic function of bounded variation on **T**. It was already known that f must be continuous. Thus f maps **T** onto a continuous, rectifiable closed curve in the plane. Arc length along the curve determines a measure on the curve. The theorem of F. and M. Riesz, in its original formulation, states that the image under f of a null set on **T** is a null set of the curve, and *vice versa*. This is equivalent to the theorem on measures of analytic type stated above, together with Theorem 17 for all functions in $\mathbf{H}^1(\mathbf{T})$.

Spaces $\mathbf{H}^p(\mathbf{T})$ can be defined also for $0 < p < 1$. The definition given above is not applicable, because only summable functions have Fourier coefficients. Instead, for these values of p, $\mathbf{H}^p(\mathbf{T})$ is defined to be the set of functions F analytic in the disk for which the integral in (1.3) is bounded as r increases to 1.

For $p \geq 1$, $\mathbf{H}^p(\mathbf{T})$ has two incarnations. As defined, it is a set of boundary functions f; or we may think of it as the corresponding set of functions F analytic in the disk. The Poisson formula provides the link between f and F. When $p < 1$, each function F of $\mathbf{H}^p(\mathbf{T})$ still has a boundary function f (Problem 4 of the next section), but the Taylor coefficients of F are not easily recovered from f, and F cannot be obtained as the Poisson integral of f. Thus for $p < 1$, $\mathbf{H}^p(\mathbf{T})$ is defined and thought of as a space of analytic functions F rather than as a space of functions on **T**.

For $p < 1$, $\mathbf{L}^p(\mathbf{T})$ is a complete metric space with distance

$$(1.7) \qquad\qquad d(f, g) = \int |f - g|^p \, d\sigma.$$

These spaces are not normed. It is not obvious how $\mathbf{H}^p(\mathbf{T})$ is embedded in $\mathbf{L}^p(\mathbf{T})$.

During the second half of the nineteenth century great progress was made in complex function theory of the plane. In this century it was recognized that analytic functions in the unit disk present problems of a different character. Sets of measure 0 on $\mathbf{T}$ arose as the sets where an analytic or harmonic function in the disk might not have radial limits. Recovering an analytic or harmonic function from its boundary function required the invention of the Lebesgue integral. This new function theory merged with new ideas about real functions and their derivatives, integration theory, approximation by trigonometric polynomials, and the spaces $\mathbf{L}^p$. Nowadays one is likely to meet the Lebesgue integral first as part of an abstract course on measure theory, separate from the classical problems of analysis. It is hard to appreciate how radical the first theorems on the boundary values of analytic functions in the disk seemed at the time, or to understand the hostility of mathematicians of the period toward such abstract and incomprehensible results.

Problems

1. Verify that (1.6) follows from the fact that the quadratic expression (1.5) has a minimum at $t = 0$.

2. Suppose that f is in $\mathbf{H}^1(\mathbf{T})$ and vanishes a.e. on an interval of $\mathbf{T}$. Show that f is null. [We shall prove more in the next section, but this can be deduced from Theorem 17.]

3. Show that trigonometric polynomials are dense in $\mathbf{L}^p(\mathbf{T})$ when $0 < p < 1$.

4. $\mathbf{H}^\infty(\mathbf{T}) + \mathbf{C}(\mathbf{T})$ is the subspace of $\mathbf{L}^\infty(\mathbf{T})$, which we shall

call **L**, consisting of all sums $f + g$ where f is in $\mathbf{H}^{\infty}(\mathbf{T})$ and g in $\mathbf{C}(\mathbf{T})$. Show that (a) **L** is closed in the uniform norm, and (b) **L** is closed under multiplication, so that it is an algebra. [This is a theorem of D. Sarason. Let $\mathbf{H}_0^{\infty}(\mathbf{T})$ be the elements of $\mathbf{H}^{\infty}(\mathbf{T})$ with mean value 0, and $\mathbf{C}_+$ the functions of $\mathbf{C}(\mathbf{T})$ that are analytic with mean value 0. The quotient space $\mathbf{C}(\mathbf{T})/\mathbf{C}_+$ has dual naturally identified with $\mathbf{H}^1(\mathbf{T})$, with the help of the theorem of F. and M. Riesz. The dual of $\mathbf{H}^1(\mathbf{T})$ is $\mathbf{L}^{\infty}(\mathbf{T})/\mathbf{H}_0^{\infty}(\mathbf{T})$. Let $\mathbf{Q}$ be the image of $\mathbf{C}(\mathbf{T})/\mathbf{C}_+$ under the natural embedding into its second dual. Then $\mathbf{Q}$ is closed. On the other hand the natural projection of $\mathbf{L}^{\infty}(\mathbf{T})$ to $\mathbf{L}^{\infty}(\mathbf{T})/\mathbf{H}_0^{\infty}(\mathbf{T})$ is continuous, and the inverse image of $\mathbf{Q}$ under this projection is **L**. Assertion (a) follows. Incidentally, Sarason has shown how to avoid the use of the Riesz theorem in the proof.]

2. Invariant subspaces; factoring; proof of the theorem of F. and M. Riesz

Let f be a nonnull function in $\mathbf{L}^2(\mathbf{T})$. The functions f, χf, $\chi^2 f, \ldots$ span a closed subspace of $\mathbf{L}^2(\mathbf{T})$ that we call $\mathbf{M}_f$. This subspace has the property that if it contains a function g, then it also contains χg. A closed subspace of $\mathbf{L}^2(\mathbf{T})$ with this property will be called an *invariant subspace*. It is not obvious whether every invariant subspace is $\mathbf{M}_f$ for some function f.

$\mathbf{H}^2(\mathbf{T})$ is an invariant subspace, generated by the constant function 1. For any f in $\mathbf{H}^2(\mathbf{T})$, $\mathbf{M}_f$ is contained in $\mathbf{H}^2(\mathbf{T})$. If $\mathbf{M}_f$ is all of $\mathbf{H}^2(\mathbf{T})$, we say that f is an *outer function*.

Let f be any bounded function. Then $\mathbf{M}_f$ is the closure of $f \cdot \mathbf{H}^2(\mathbf{T})$, the set of all products fg where g belongs to $\mathbf{H}^2(\mathbf{T})$. If f is bounded and bounded from 0, this set of products is closed and equals M_f.

A. Beurling characterized the invariant subspaces that are contained in $H^2(T)$. His theorem has been generalized to this result.

Theorem 18. *Each invariant subspace* **M** *of* $L^2(T)$ *belongs to one of two types. Either it consists exactly of all functions of* $L^2(T)$ *that are supported on some fixed measurable subset of* **T**, *or it is* $q \cdot H^2(T)$ *for some function* q *of modulus* 1 *a.e.* (The first kind of subspace is a *Wiener subspace*, the second a *Beurling subspace*.)

By hypothesis, $\chi \cdot M$ is contained in **M**. Suppose that these subspaces are the same. That is, **M** is invariant under multiplication by χ^{-1} as well as by χ; hence it is invariant under multiplication by χ^n for every integer n. Let f be any element of **M**, and g of the orthogonal complement of **M**. Since $f\chi^n$ is in **M** for every n,

$$(2.1) \qquad \int f\bar{g}\chi^n \, d\sigma = 0 \qquad \text{(all } n\text{)}.$$

Thus the summable function $f\bar{g}$ has all its Fourier coefficients 0, and so equals 0 a.e. That is, the supports of f and of g are disjoint.

Let w be the supremum of the measures of sets E in **T** with the property that some function of **M** is different from 0 a.e. in E. Find a sequence (E_n) of such sets whose measures tend to w, and denote by E their union. By what has just been shown, any g in the orthogonal complement of **M** must vanish a.e. on each E_n, and therefore a.e. on E. A function f of $L^2(T)$ that is supported on E is orthogonal to such g, and therefore belongs to **M**.

Now let f be any element of **M**. If f is not supported on E, we could add to f some function supported on E (which belongs to M) to obtain a function in M whose support has measure greater than the measure of E. This is impossible. Therefore all the elements of M are supported on E, and M is the Wiener subspace consisting of all functions supported on E.

Otherwise $\chi \cdot M$ is a proper subset of M. Find a function q of norm 1 in M that is orthogonal to $\chi \cdot M$. Then q is orthogonal to $\chi^n q$ for positive integers n:

$$(2.2) \qquad \int |q|^2 \chi^n \, d\sigma = 0 \qquad (n \geq 1).$$

Taking the complex conjugate of this equality shows that it holds also for n negative. This means that the Fourier coefficients of $|q|^2$ are 0 except for its mean value, so $|q|^2$ is constant. The constant is 1 because q has norm 1.

The functions $\chi^n q$ form an orthonormal basis for $\mathbf{L}^2(\mathbf{T})$. For $n \geq 0$, these functions belong to M. But $\chi^{-n} q$ is orthogonal to M for $n > 0$ (this is the same as the fact that q is orthogonal to $\chi^n \cdot M$). It follows that M equals the closed span of $(\chi^n q)$ $(n \geq 0)$, which is $q \cdot \mathbf{H}^2(\mathbf{T})$. This completes the proof.

Beurling's proof was function-theoretic (and required that M be a subset of $\mathbf{H}^2(\mathbf{T})$), but in the other direction, Beurling showed that the result could provide function-theoretic information. This proof in Hilbert space gives a new way to derive function-theoretic results, and to generalize them to other contexts.

For example, Theorem 18 contains Theorem 17. Let f be any nonnull function in $\mathbf{H}^2(\mathbf{T})$. Since M_f is contained in $\mathbf{H}^2(\mathbf{T})$ it cannot be a Wiener subspace, and must have the form $q \cdot \mathbf{H}^2(\mathbf{T})$

for some unimodular function q. If f were to vanish on a set of positive measure, the same would be true of all the functions of $\mathbf{M}_f$, in particular of q, but this is impossible.

A unimodular function in $\mathbf{H}^2(\mathbf{T})$ is called *inner*. Let f be any nonnull function in $\mathbf{H}^2(\mathbf{T})$, with $\mathbf{M}_f = q \cdot \mathbf{H}^2(\mathbf{T})$ where q has modulus 1 a.e. We have just observed that q belongs to $\mathbf{M}_f$ and thus to $\mathbf{H}^2(\mathbf{T})$, so q is inner. Moreover $f = qg$ for some g in $\mathbf{H}^2(\mathbf{T})$, and this function g is outer:

$$(2.3) \qquad \mathbf{M}_g = \mathbf{M}_{\overline{q}f} = \overline{q} \cdot \mathbf{M}_f = \mathbf{H}^2(\mathbf{T}).$$

Every nonnull element of $\mathbf{H}^2(\mathbf{T})$ *is the product of an inner and an outer function.*

The factoring is unique, aside from constant factors of modulus 1. For suppose that $qg = pk$ where p and q are inner, g and k outer. Then $\overline{p}qg = k$ is outer, and

$$(2.4) \qquad \mathbf{H}^2(\mathbf{T}) = \mathbf{M}_{\overline{p}qg} = \overline{p}q \cdot \mathbf{M}_g = \overline{p}q \cdot \mathbf{H}^2(\mathbf{T}).$$

This implies that $\overline{p}q$ and $p\overline{q}$ are both in $\mathbf{H}^2(\mathbf{T})$. Hence the product is constant, as we wished to show.

We have proved

Theorem 19. *Every nonnull function in* $\mathbf{H}^2(\mathbf{T})$ *can be factored as the product of an inner and an outer function. The factoring is unique aside from constant factors of modulus 1.*

This factoring theorem can be extended to $\mathbf{H}^1(\mathbf{T})$. Let f be a nonnull element of the space. Set $w = |f|^{1/2}$, belonging to $\mathbf{L}^2(\mathbf{T})$.

Lemma. $\mathbf{M}_w$ *is a Beurling subspace.*

Set $p = f/|f|$ where $f \neq 0$, $p = 0$ where $f = 0$. Then $f = w^2 p$.

For $n \geq 1$,

$$(2.5) \qquad 0 = \int f\chi^n \, d\sigma = \int (wp)(w\chi^n) \, d\sigma.$$

That is, $w\bar{p}$ is orthogonal to $\chi^n w$ for such n. If $\mathbf{M}_w$ were a Wiener subspace, this would imply that $w\bar{p}$ and and w have disjoint supports, so that $w = 0$, contrary to assumption.

Thus $\mathbf{M}_w = q \cdot \mathbf{H}^2(\mathbf{T})$ for some unimodular function q. It follows that $w = qg$ for some g that is outer by the calculation of (2.3), and $f = w^2 p = pq^2 g^2$. We want to show that pq^2 is inner.

Choose a sequence of analytic trigonometric polynomials (P_n) such that $P_n g$ tends to the constant function 1 in $\mathbf{H}^2(\mathbf{T})$. This is possible because g is outer. Then

$$(2.6) \qquad P_n f = pq^2 (P_n g) g$$

converges in $\mathbf{L}^1(\mathbf{T})$ to $pq^2 g$, which thus belongs to $\mathbf{H}^1(\mathbf{T})$. But this function is in $\mathbf{L}^2(\mathbf{T})$, so it is in $\mathbf{H}^2(\mathbf{T})$. Multiply by P_n again and take the limit in $\mathbf{L}^2(\mathbf{T})$; we obtain pq^2, which therefore is in $\mathbf{H}^2(\mathbf{T})$ as we wished to show.

Theorem 20. *Every nonnull function in* $\mathbf{H}^1(\mathbf{T})$ *can be factored in the form* qg^2 *where* q *is inner and* g *is outer in* $\mathbf{H}^2(\mathbf{T})$.

As a corollary we obtain this improvement of Theorem 17, also due to F. and M. Riesz: *a function in* $\mathbf{H}^1(\mathbf{T})$ *that vanishes on a set of positive measure is identically* 0.

This interesting proposition is also nearly proved.

Corollary. *If the set* E *in the open unit disk is the set of zeros of a nonnull function in* $\mathbf{H}^1(\mathbf{T})$, *then* E *is the set of zeros of an inner function.*

Since $\mathbf{H}^1(\mathbf{T})$ contains the other spaces $\mathbf{H}^p(\mathbf{T})$ $(p > 1)$, the corollary means that the zero sets of the functions in the various spaces $\mathbf{H}^p(\mathbf{T})$ are the same for all $p \geq 1$.

The result is obvious if we prove this statement: *an outer function of $\mathbf{H}^2(\mathbf{T})$ has no zeros inside the unit disk.*

For each z, the functional that maps f to $F(z)$ is a continuous linear functional on $\mathbf{H}^2(\mathbf{T})$. Let g be an outer function of $\mathbf{H}^2(\mathbf{T})$, and G its analytic extension to the disk. If $G(z) = 0$ for some z, then this functional would vanish on all the functions of $\mathbf{M}_g$, which therefore cannot be all of $\mathbf{H}^2(\mathbf{T})$.

It would be reasonable to think that, conversely, f is outer if $F(z)$ never vanishes. This would be the case if every non-constant inner function had a zero in the disk. But this is not true; later we shall find inner functions without zeros.

Finally we prove the theorem of F. and M. Riesz. An analytic measure μ generates the analytic function

$$(2.7) \qquad F(z) = \sum_0^\infty \hat{\mu}(n)\, z^n,$$

which satisfies

$$(2.8) \qquad \int |F(re^{ix})|\, d\sigma(x) \leq \|\mu\| \qquad (0 < r < 1).$$

Suppose, for the moment, that all the zeros of F in the disk are incorporated in an inner function Q. Then $G = F/Q$ is analytic in the disk, and it is true (although not obvious) that G satisfies the same condition (2.8). Since it has no zeros in the disk, G has an analytic square root H, which belongs to $\mathbf{H}^2(\mathbf{T})$ in the function-theoretic sense of (1.3). But the square of a function in $\mathbf{H}^2(\mathbf{T})$ has a boundary function that belongs to $\mathbf{H^1(T)}$ in sense of Fourier coefficients, and $F = QH^2$ is the Poisson integral of its boundary function qg. This is the conclusion of the theorem.

In the next sections we shall get the, function-theoretic means for making a full proof out of this idea, but the following proof is more direct.

As usual let f_r be the function with values $F(re^{ix})$. By Theorem 19, $f_r = q^r g^r$ for some inner q^r and outer g^r. (The superscripts are indices, not exponents.) Since $G^r(z)$ is never 0, it has an analytic square root H^r. From (2.8) we see that $||h^r||_2$ is bounded for $0 < r < 1$. Let h be an element of $\mathbf{H^2(T)}$ that is the weak limit of (h^r) as r increases to 1 through some sequence of values. Since evaluation at an interior point of the disk is a continuous linear functional on $\mathbf{H^2(T)}$, $H(z) = \lim H^r(z)$ for each z in the disk, when r increases to 1 through that sequence. The equality $f_r = q^r g^r$ gives $F(rz) = Q^r(z)G^r(z)$ for $|z| < 1$, so that

$$(2.9) \qquad F(z) \; = \; \lim_r F(rz) \; = \; \lim_r Q^r(z)(H^r(z))^2.$$

The factor $Q^r(z)$ has modulus bounded by 1, and the square on the right converges for each z in the disk to $H(z)^2$. By Hurwitz' theorem $H(z)$ is never 0; for our purpose it suffices to note the obvious fact that it is not identically 0. Therefore $Q^r(z)$ converges to a bounded analytic function Q. Thus we have a factoring $F = QH^2$ where Q is bounded and H is in $\mathbf{H^2(T)}$.

The analytic function QH^2 is the Poisson integral of its boundary function qh^2, because the product belongs to the space $\mathbf{H^1(T)}$ (defined by Fourier coefficients). Thus F is the Poisson integral of a summable boundary function, which by the unicity theorem has to be the measure μ. This completes the proof.

Problems

1. Justify the assertion, used above, that $\mathbf{M}_{qf} = q \cdot \mathbf{M}_f$ for q a unimodular function and f in $\mathbf{L}^2(\mathbf{T})$.

2. Show that the product of two functions in $\mathbf{H}^2(\mathbf{T})$ is in $\mathbf{H}^1(\mathbf{T})$. (This fact was used above.)

3. Let g be a bounded analytic function on $\mathbf{T}$. Show that e^g is outer.

4. Imitate the proof above to show that every function in $\mathbf{H}^{1/2}(\mathbf{T})$ is the product (inside the disk) of two functions of $\mathbf{H}^1(\mathbf{T})$. Deduce that functions of $\mathbf{H}^{1/2}(\mathbf{T})$ have radial limits a.e.; and if the boundary function vanishes on a set of positive measure, the function is null. Show that the boundary function is in $\mathbf{L}^{1/2}(\mathbf{T})$. (Thus $\mathbf{H}^{1/2}(\mathbf{T})$ is identified with a subspace of $\mathbf{L}^{1/2}(\mathbf{T})$.) Show that $\mathbf{H}^{1/2}(\mathbf{T})$ is closed in $\mathbf{L}^{1/2}(\mathbf{T})$. Extend the results to $\mathbf{H}^p(\mathbf{T})$ for $p > 0$.

5. Show that if F is in $\mathbf{H}^p(\mathbf{T})$, $p < 1$, and its boundary function is summable, then F is in $\mathbf{H}^1(\mathbf{T})$.

6. Show that f in $\mathbf{H}^2(\mathbf{T})$ is outer if and only if it has this property: if h is in $\mathbf{L}^\infty(\mathbf{T})$ and fh is in $\mathbf{H}^2(\mathbf{T})$, then h is analytic.

7. Show that if g and h are outer functions in $\mathbf{H}^2(\mathbf{T})$, then gh is outer in $\mathbf{H}^1(\mathbf{T})$, in the sense that the smallest closed subspace of $\mathbf{H}^1(\mathbf{T})$ containing $\chi^n gh$ for all $n \geq 0$ is $\mathbf{H}^1(\mathbf{T})$ itself.

8. Prove the lemma of Fejér and Riesz: *every non-negative trigonometric polynomial ϕ is $|\psi|^2$ for some trigonometric polynomial ψ.* [Let n be the smallest integer such that $\chi^n \phi$ is analytic. Write $\phi = |\psi|^2$ with ψ outer in $\mathbf{H}^2(\mathbf{T})$. Deduce that ψ is a trigonometric polynomial from the fact that $\psi \chi^n \overline{\psi}$ is analytic.]

3. Theorems of Szegő and Beurling

The preceding theorems do not tell which functions in $\mathbf{H}^2(\mathbf{T})$ are outer. Theorem 17 says that a nonnull analytic function cannot vanish on a set of positive measure, but does not tell what the modulus of such a function can be. These questions can be answered. Beurling characterized outer functions; and older theorems of Szegő describe the modulus of functions of $\mathbf{H}^p(\mathbf{T})$ for all p. Beurling did not notice that his theorem can be deduced from Szegő's results. Before stating and proving these theorems we develop some needed function theory.

Let w be a function on $\mathbf{T}$ satifying $0 \le w \le 1$. Since $\log w < w$, $\log w$ is summable above but may fail to be summable below. Assume that $\log w$ is summable, and call its Fourier coefficients (a_n). Form the analytic function

$$(3.1) \qquad H(z) \;=\; a_0 + 2\sum_1^\infty a_n z^n.$$

The real part of H is the negative harmonic function $P_r * \log w$.

Lemma 1. $F = \exp H$ *is a bounded analytic function on the disk whose boundary function f satisfies $|f| = w$ a.e.*

Since $\log w$ is negative, $P_r * \log w$ is also negative and thus

$$(3.2) \qquad |\exp H| \;=\; \exp \Re H \le 1$$

throughout the disk. As r increases to 1, $\exp(P_r * \log w)$ tends to $\exp \log w$ almost everywhere. This proves the lemma.

Lemma 2. f *is outer.*

Factor f by Theorem 19: $f = qg$ with q inner and g outer. If f is not outer, q is not constant, so its analytic extension Q has

$|Q(0)| < 1$. Q has no zeros, because F has none. Thus

$$\log|F(0)| = \int \log|f|\, d\sigma = \int \log|g|\, d\sigma \geq$$

(3.3) $$\limsup_{r \uparrow 1} \int \log|G(re^{ix})|\, d\sigma(x) = \log|G(0)| >$$

$$\log|Q(0)G(0)| = \log|F(0)|.$$

The first inequality is Fatou's theorem, where the sense is opposite to the usual one because the integrand is non-positive. The contradiction proves the lemma.

The lemmas give a way to construct outer functions with given modulus. We want to show that the construction gives *all* outer functions. For the moment we restrict ourselves to bounded functions.

Lemma 3. *Let f be a bounded outer function, with analytic extension F. Then $\log|F(z)|$ is harmonic in the disk, and is the Poisson integral of its boundary function $\log|f|$.*

Say $|f| \leq 1$. We know that F has no zeros in the disk, so $\log|F|$ is a negative harmonic function. We have

(3.4) $$\int |\log|F(re^{ix})||\, d\sigma(x) = -\int \log|F(re^{ix})|\, d\sigma(x) = -\log|F(0)|$$

for each $r < 1$. By Theorem 3 of Chapter 1, Section 6, $\log|F|$ is the Poisson integral of some measure μ on $\mathbf{T}$, which here is necessarily a negative measure. We want to show that μ is the absolutely continuous measure $\log|f|\, d\sigma$.

From Theorem 4 and Problem 4, both of Chapter 1, Section 7, the radial limits of $\log|F(z)|$ are exactly the absolutely continuous part of $d\mu$; the lemma will be proved if we show that the singular part of μ vanishes.

Suppose the singular part μ_s of μ is not null. At any rate it is a negative measure. Form an analytic function H' by (3.1), using the Fourier-Stieltjes coefficients of μ_s. The associated function $F' = \exp H'$ is bounded again. Call its boundary function f'. The real part of H' has radial limits 0 a.e., so that $|f'| = 1$ a.e. and f' is a non-constant inner function.

F/F' is analytic in the disk, and $\log|F/F'|$ is the Poisson integral of $\log|f|$. Therefore the quotient is bounded in the disk, which means that f is divisible in $\mathbf{H}^\infty(\mathbf{T})$ by the inner function f'. This contradicts the assumption that f was outer, and the lemma is proved.

Beurling's Theorem. f in $\mathbf{H}^2(\mathbf{T})$ *is outer if and only if*

$$(3.5) \qquad \int \log|f|\, d\sigma = \log|F(0)| > -\infty.$$

Given that $F(z)$ is never 0 in the disk, a necessary condition for f to be outer, the formula expresses the fact that the harmonic function $\log|F(z)|$ has the mean value property: its value at the origin is the mean value of its values on $\mathbf{T}$. The formula is also the Poisson representation of $\log|F(0)|$ in terms of its boundary function. At first it seems that (3.5) should always be true, and it certainly is if F is sufficiently smooth near the boundary. However, from the fact that F is the Poisson integral of its boundary function f, it does not follow that the harmonic function $\log|F(z)|$ is the Poisson integral of its boundary function $\log|f|$. For example if f is a non-constant inner function, even one without zeros, the left side of (3.5) is 0, but the right side is negative.

First suppose that f is bounded. If f is outer, then we have shown above that $\log|F(z)|$ is the Poisson integral of its boundary function $\log|f|$, and (3.5) is a special case of this representation. If f is not outer, then $f = qg$ with q a non-constant inner function and g outer. We have

$$(3.6) \qquad \int \log|f|\,d\sigma = \int \log|g|\,d\sigma = \log|G(z)| >$$
$$\log|Q(0)\,G(0)| = \log|F(0)|.$$

This proves the theorem for bounded functions.

If f is not bounded let $w^{-1} = \max(|f|, 1)$. Then $\log w$ is non-positive and summable (indeed to every positive power). Form the analytic function H as in (3.1) with the Fourier coefficients of $\log w$, and let $K = \exp H$, with boundary function k. Then k is a bounded outer function by Lemma 2, and kf is in $\mathbf{H}^\infty(\mathbf{T})$.

Lemma 4. *f is outer if and only if kf is outer.*

The simple proof is asked for below.

Being outer and bounded, k satisfies Beurling's condition (3.5). Therefore kf satisfies the condition if and only if f does. This remark and Lemma 4 complete the proof of Beurling's theorem.

Corollary. *For any f in $\mathbf{H}^1(\mathbf{T})$ with analytic extension F*

$$(3.7) \qquad \int \log|f|\,d\sigma \geq \log|F(0)|.$$

If f is sufficiently regular then (3.7), with equality and some supplementary terms to take account of zeros of F in the disk, is a classical theorem of function theory called *Jensen's equality*. The name *Jensen's inequality* was invented for (3.7) by David Lowdenslager and the author.

Write $f = qg^2$ where q is inner and g is outer in $\mathbf{H}^2(\mathbf{T})$ (by Theorem 20 of the last section). If q is constant (3.7) is equality; otherwise we find inequality as in (3.6).

Corollary. *For any nonnull function f in $\mathbf{H}^1(\mathbf{T})$, $\log|f|$ is summable.*

After division of f by a power of χ if necessary, $F(0) \neq 0$, so the right side of (3.7) is finite.

Szegö's Theorem. *Let w be a non-negative summable function on $\mathbf{T}$. Then*

$$(3.8) \qquad \exp \int \log w \, d\sigma = \inf_P \int |1 + P|^2 \, w \, d\sigma$$

where P ranges over all analytic trigonometric polynomials with mean value 0. The right side is 0 if $\log w$ is not summable.

Of course $\log w$ is summable above; but if w is too small, for example if w vanishes on a set of positive measure, $\log w$ is not summable below.

Suppose that $w = |f|^2$ for some f in $\mathbf{H}^2(\mathbf{T})$. f can be taken to be outer by Theorem 19. For each P of the form envisaged in the theorem, $(1 + P)f$ is analytic with mean value $a_0(f)$; and because f is outer, such products can approximate the constant function $a_0(f)$ in the norm of $\mathbf{H}^2(\mathbf{T})$. Thus

$$(3.9) \qquad \inf \int |1 + P|^2 \, w \, d\sigma = \inf \int |(1 + P)f|^2 \, d\sigma \leq$$
$$|a_0(f)|^2 = \exp \int \log |f|^2 \, d\sigma.$$

The last equality is (3.5). But for every P,

$$(3.10) \qquad \int |(1 + P)f|^2 \, d\sigma \geq |a_0(f)|^2$$

by the Parseval relation. This proves the theorem if w has the suggested form.

Lemma 1 shows that w is such a square if it is bounded. Otherwise set $r = \min(w, 1)$, $s = \max(w, 1)$. There are outer functions g, h such that $|g|^2 = r$, $|h|^2 = 1/s$. Set $f = g/h$. Then $|f|^2 = w$. We must show that f is in $\mathbf{H}^2(\mathbf{T})$.

We have $g = fh$. Choose analytic trigonometric polynomials P_n such that $P_n h$ tends to the constant function 1 in the norm of $\mathbf{H}^2(\mathbf{T})$. Then $P_n g = f P_n h$ converges to f in the norm of $\mathbf{L}^1(\mathbf{T})$. Since the functions $P_n g$ are analytic, f is in $\mathbf{H}^1(\mathbf{T})$, and being square-summable, belongs to $\mathbf{H}^2(\mathbf{T})$. This finishes the proof for the case that w and $\log w$ are summable.

If $\log w$ is not summable, we are to show that the right side of (3.8) is 0. Set $w_n = \max(w, 1/n)$. The formula holds for each w_n:

$$(3.11) \quad \inf \int |1 + P|^2 \, w \, d\sigma \leq \inf \int |1 + P|^2 \, w_n \, d\sigma = \exp \int \log w_n \, d\sigma.$$

The right side tends to 0, so the left side must equal 0. This completes the proof.

On the way we have proved this important fact: *each non-negative summable function w such that $\log w$ is summable is the modulus of a function in $\mathbf{H}^1(\mathbf{T})$.*

Problems

1. Prove Beurling's theorem from Szegö's theorem.

2. A function f in $\mathbf{H}^1(\mathbf{T})$ is called outer if the functions $(\chi^n f)$ $(n \geq 0)$ span a dense subspace of $\mathbf{H}^1(\mathbf{T})$. Show that f is outer if and only if it satisfies (3.5).

3. Prove Lemma 4.

4. Show that if f is outer in $\mathbf{H}^1(\mathbf{T})$, then $\log|F|$ is the Poisson integral of its boundary function $\log|f|$.

5. Let G be analytic in the unit disk and suppose that $F = \exp G$ is in $\mathbf{H}^1(\mathbf{T})$. Show that F is outer, and that all outer functions of $\mathbf{H}^1(\mathbf{T})$ are obtained in this way.

4. Structure of inner functions

The last section gave a function-theoretic description of outer functions. This section treats inner functions.

A *Blaschke factor* is a function $(z - z_0)/(1 - \bar{z}_0 z)$, where z_0 is a point in the interior of the unit disk. This is a fractional linear transformation that maps the disk onto itself. A product of such factors maps the disk into itself, and the circle $\mathbf{T}$ onto itself. A finite product of Blaschke factors is an inner function. Our first step is to obtain infinite products.

Let q be an inner function with analytic extension Q. Suppose that Q has infinitely many zeros $z_1, z_2, \ldots$. Zeros can have multiplicity greater than 1; each zero is repeated in the sequence according to its multiplicity. We assume that $Q(0) \neq 0$. If F_n is the product of the Blaschke factors formed with $z_1, \ldots, z_n$, then Q/F_n is still a bounded function in the disk, and therefore is inner. By Jensen's inequality

$$(4.1) \quad 0 \geq \log|Q(0)| - \log|F_n(0)| = \log|Q(0)| - \sum_1^n \log|z_j|.$$

This is true for any n; therefore

$$(4.2) \qquad \sum_1^\infty \log 1/|z_j| \leq \log \frac{1}{|Q(0)|} < \infty.$$

We knew that $|z_j|$ tends to 1 (if Q has infinitely many zeros), because the zeros cannot accumulate at an interior point of the disk; (4.2) is a quantitative refinement of this fact. If $Q(0) = 0$, the sum (4.2) over all zeros z_j of Q except 0 is still finite. The Corollary of Theorem 20 shows that the same result holds for all nonnull functions of $\mathbf{H^1(T)}$.

The finiteness of the sum in (4.2) is called the *Blaschke condition* for a set of points in the disk. It is equivalent to

$$(4.3) \qquad \sum_1^\infty (1-|z_j|) < \infty \quad \text{or} \quad \prod_1^\infty |z_j| > 0.$$

We want to show that every set of points (z_j) in the disk satisfying the Blaschke condition is the set of zeros of an inner function. (It is still assumed that no z_j is 0.)

For each n, let

$$(4.4) \qquad B_n(z) = \prod_1^n \frac{z-z_j}{1-\bar{z}_j z} \frac{|z_j|}{-z_j};$$

the expression attached to each Blaschke factor on the right has modulus 1 and makes it positive at the origin. Consequently

$$(4.5) \qquad \lim B_n(0) = \prod_1^\infty |z_j| > 0.$$

Thus (B_n) is a bounded family of functions on the disk converging at the origin to a nonzero limit. We want to show that the sequence converges uniformly on compact subsets of the open disk to a limit function that is inner with zeros at the points (z_j).

If the sequence converges as asserted, the limit function B is certainly 0 at each z_j, and by the theorem of Hurwitz, nowhere else. Moreover each zero of B has the proper multiplicity.

A sequence of complex numbers (w_n) has convergent product if and only if no w_n is 0 and

$$(4.6) \qquad\qquad \sum_1^\infty |1 - w_n| < \infty.$$

A calculation shows that the factors w_n of (4.4) satisfy this condition for each z, $|z| < 1$. Hence $B_n(z)$ converges throughout the disk, and the limit function has modulus bounded by 1.

The fact that the convergence is uniform on compact subsets of the disk follows from pointwise convergence and the uniform boundedness of the sequence. (By the Lebesgue convergence theorem, $B_n(re^{ix})$ converges in $\mathbf{L}^1(\mathbf{T})$ to $B(re^{ix})$ for each $r < 1$. Cauchy's theorem now gives uniform convergence inside any circle about the origin of radius smaller than r.) Thus B is an analytic function bounded in the disk.

It is not yet proved that B is inner, that is, that its boundary function b has modulus 1 a.e. Factor b as qg, where q is inner and g is outer. Their analytic extensions satisfy $B = QG$ in the disk. Since $|g| = |b|$, g and therefore also G have modulus bounded by 1. Write

$$(4.7) \qquad\qquad \frac{B}{B_n} = \left(\frac{Q}{B_n}\right) G.$$

The function on the left tends pointwise to 1 inside the disk. On the right the first factor is the ratio of inner functions, and is bounded, so it is inner and

$$(4.8) \qquad\qquad \left|\frac{Q}{B_n}\right| \leq 1.$$

If $|G(z)| < 1$ for any z in the disk, (4.7) would be contradicted as n tends to ∞. Thus $|G(z)| = 1$ for all z, G is constant, and b is an inner function.

A *Blaschke product* is a finite or infinite product

$$(4.9) \qquad B(z) = z^k \prod \frac{z - z_j}{1 - \overline{z}_j z} \frac{|z_j|}{-z_j}$$

where the sequence (z_j) contained in the open unit disk satisfies (4.3), and k is a non-negative integer. No z_j is 0, but the product may, after all, vanish at the origin.

Let q be any inner function whose analytic extension has zeros (z_j), repeated according to multiplicity. Form the Blaschke product b on these zeros. Then $s = q/b$ is inner, because Q/B_n is inner for each n and converges to $S = Q/B$. The function S has no zeros; s is a *singular* inner function. Then $-\log|S(z)|$ is a positive harmonic function, the Poisson integral of a positive measure μ on **T**. Theorem 4 of Chapter 1, Section 7, together with Problem 4 of that section, shows that μ is a singular measure.

The Poisson kernel can be written

$$(4.10) \qquad P_r(e^{ix}) = \Re\left(\frac{1 + re^{ix}}{1 - re^{ix}}\right).$$

Therefore

$$(4.11) \qquad F(z) = -\int \frac{1 + re^{i(x-t)}}{1 - re^{i(x-t)}} \, d\mu(t) \qquad (z = re^{ix})$$

is an analytic function in the disk whose real part equals $\log|S(z)|$. That is,

(4.12) $|S(z)| = |\exp F(z)|.$

After adjusting a constant factor of modulus 1,

(4.13) $S(z) = \exp - \int \dfrac{e^{it} + z}{e^{it} - z} \, d\mu(t).$

Every inner function is the product of a (possibly empty) Blaschke product and a (possibly constant) singular function. These factors are determined up to a multiplicative factor of modulus 1 by a set of points in the disk satisfying Blaschke's condition, and a positive singular measure on **T**. This is a complete function-theoretic description of inner functions, and thus of invariant subspaces of $\mathbf{H}^2(\mathbf{T})$.

Problems

1. Establish the equivalence of the three forms of the Blaschke condition, embodied in (4.2) and (4.3).

2. Let **M** be an invariant subspace properly contained in $\mathbf{H}^2(\mathbf{T})$. Show that there is an invariant subspace **N** properly between **M** and $\mathbf{H}^2(\mathbf{T})$, unless **M** has a particular simple form. Determine this form.

3. Show that every nonnull invariant subspace of $\mathbf{L}^2(\mathbf{T})$ properly contains another nonnull invariant subspace.

4. Show that if f and $1/f$ are both in $\mathbf{H}^1(\mathbf{T})$, then f is outer. If f is in $\mathbf{H}^1(\mathbf{T})$ and $\Re f \geq 0$, then f is outer.

5. Show that every continuous function on **T** with modulus 1 can be approximated uniformly by functions p/q, where p and q are finite Blaschke products. [This problem is hard.]

6. Let q be inner with $Q(0) = 0$. As a map of the circle to

itself, q carries the measure σ to a new measure ν on $\mathbf{T}$: $\nu(E) = \sigma(e^{ix}: q(e^{ix})$ is in $E)$. Show that $\nu = \sigma$. [For any continuous function h on $\mathbf{T}$, $\int h(e^{ix})\,d\nu(x) = \int h \circ q(e^{ix})\,d\sigma(x)$. Calculate the Fourier-Stieltjes coefficients of ν.]

5. Theorem of Hardy and Littlewood; Hilbert's inequality

Fourier coefficients of summable functions do not in general satisfy any condition of smallness except that they tend to 0 (Mercer's theorem). Therefore this theorem about one-sided series is unexpected.

Theorem of Hardy and Littlewood. *For all h in $\mathbf{H}^1(\mathbf{T})$*

$$(5.1) \qquad \sum_0^\infty |a_n(h)|/(n+1) \le \pi \|h\|_1.$$

Factor h by Theorem 20: $h = qg^2$, where q is inner and g is outer in $\mathbf{H}^2(\mathbf{T})$. Let g have Fourier coefficients (b_n), and qg have coefficients (c_n). Then

$$(5.2) \qquad a_n(h) = \sum_{k=0}^n b_k c_{n-k} \qquad (n \ge 0).$$

Hence

$$(5.3) \qquad \sum_0^\infty \frac{|a_n(h)|}{n+1} = \sum_{n=0}^\infty \frac{1}{n+1}\Big|\sum_{k=0}^n b_k c_{n-k}\Big| \le \sum_{j,k=0}^\infty \frac{|b_j c_k|}{j+k+1}.$$

Moreover $\|h\|_1 = \|g\|_2 \|qg\|_2$. Therefore the theorem is a consequence of the

Double series theorem of Hilbert. *For any square-summable sequences (b_k), (c_k) $(k = 0, 1, \ldots)$*

(5.4)
$$\sum_{j,k=0}^{\infty} \frac{|b_j c_k|}{j+k+1} \leq \pi ||b||_2 ||c||_2.$$

The constant π is best possible.

The constant was determined by I. Schur.

Before proceeding further we introduce the notions of *bilinear* and *Hermitian* forms, perhaps familiar from linear algebra. Let $B(j, k)$ be any function of two integer variables. The bilinear form with kernel B is the formal sum

(5.5)
$$B = \sum_{j,k=-\infty}^{\infty} B(j, k) \, x_j y_k.$$

The sum has meaning if x_j, y_k are 0 except for finitely many indices, and is linear in each of the vectors $X = (x_j)$, $Y = (y_k)$. The Hermitian form with the same kernel is

(5.6)
$$H = \sum_{j,k=-\infty}^{\infty} B(j, k) \, x_j \bar{y}_k,$$

linear in X but conjugate linear in Y.

The form B is *bounded* in l^2 if there is a number K such that

(5.7)
$$|B| \leq K ||X||_2 ||Y||_2$$

for all X and Y ending in zeros. The smallest such K is the *bound* of B. The same definitions are made for H, and the bounds are obviously the same.

If B is bounded, (5.5) defines a continuous function on a dense subset of $l^2 \times l^2$, which has a unique continuous extension to the whole space. The extended function B also satisfies (5.7).

A transformation S of sequences is defined by setting

(5.8) $(SX)_k = \sum_j B(j, k)x_j$

for sequences ending in zeros. If B is bounded, (5.7) implies that the bound of S is at most K. Hence S has a unique continuous extension to all of $\mathbf{l}^2$, which we still call S. The same is true about H.

It is usually difficult to decide whether a kernel B determines a bounded form. Two applications of the Schwarz inequality to (5.5) give

(5.9) $|B| \leq \left(\sum_{j,k} |B(j, k)|^2\right)^{1/2} ||X||_2 ||Y||_2,$

so the finiteness of the sum certainly implies that the bilinear (or Hermitian) form is bounded. However, this condition is too strong; it is not satisfied by the kernel in (5.4), for example.

Hilbert's form has coefficients $B(j, k) = (j + k + 1)^{-1}$ for j, $k \geq 0$, and $= 0$ for other j, k. Considered merely on the set j, $k \geq 0$, the coefficients have the special form $B(j, k) = B'(j + k)$. For this reason (5.8) can be rewritten as a convolution, so the problem can be restated and solved on the dual group.

Define $\phi(e^{ix}) = x - \pi$ on $(0, 2\pi)$. Then $a_0(\phi) = 0$, $a_n(\phi) = i/n$ for other n. Let f and g be trigonometric polynomials, which we write in the unusual way

(5.10) $f = \sum b_j \chi^{-j}, \qquad g = \sum c_k \chi^{-k}.$

Then

$$(5.11) \qquad \int fg\phi\chi^{-1}\, d\sigma = \sum_{j,k} b_j c_k a_{j+k+1}(\phi).$$

Since ϕ is bounded by π, the modulus of the left side is less than

$$(5.12) \qquad \pi\|f\|_2\|g\|_2 = \pi\|b\|_2\|c\|_2.$$

If b_j and c_k are 0 for j, $k < 0$, we find

$$(5.13) \qquad \left| \sum_{j,k=0}^{\infty} \frac{b_j c_k}{j+k+1} \right| \le \pi\|b\|_2\|c\|_2.$$

The right side is unchanged if b_j, c_k are replaced by their moduli, so that (5.4) is proved under the assumption that the sum has only finitely many terms. The passage to the limit is easy.

The proof that π is the best constant is Problem 4 below.

The same proof shows that an inequality

$$(5.14) \qquad \left| \sum_{j,k \ge 0} B(j+k)\, b_j c_k \right| \le K\|b\|_2\|c\|_2$$

holds if $B(n)$ equals $a_n(\phi)$ for non-negative n, where ϕ is a function with uniform bound K. If the values of B are non-negative, then b_j, c_k can be replaced by their moduli.

The theorem of Hardy and Littlewood has an interesting reformulation: *Let f be a function of bounded variation on* **T** *with* $a_n(f) = 0$ *for* $n < 0$. *Then the Fourier series of f converges absolutely, and*

$$(5.15) \qquad \sum_{1}^{\infty} |a_n(f)| \le V/2$$

where V is the total variation of f.

We may assume (by rotating f) that f is continuous at 1; and (by adding a constant) that $f(1) = 0$. Let μ be the measure associated with f; at points of continuity of f

$$(5.16) \qquad f(e^{ix}) = \int\limits_0^x d\mu(t).$$

Integration by parts (see the Appendix) gives for the Fourier-Stieltjes coefficints of μ

$$(5.17) \qquad \int \chi^{-n} d\mu = in \int e^{-inx} f(e^{ix})\, dx = 2\pi i n a_n(f).$$

Thus μ is a measure of analytic type. By the theorem of F. and M. Riesz (Section 1), μ is absolutely continuous. Thus f is continuous, differentiable almost everywhere, $g(e^{ix}) = df(e^{ix})/dx$ is in $\mathbf{H}^1(\mathbf{T})$, and $a_n(g) = in a_n(f)$ for all n.

In particular $a_0(g) = 0$. If we apply the theorem of Hardy and Littlewood to $e^{-ix} g(e^{ix})$ we find

$$(5.18) \qquad \sum_0^\infty |a_{n+1}(f)| \le \pi \|g\|_1 = V/2.$$

This completes the proof.

Problems

1. Prove another inequality of Hilbert:

$$\left| \sum_{j \ne k} \frac{b_j c_k}{j - k} \right| \le \pi \|b\|_2 \|c\|_2,$$

valid for b, c in $\mathbf{l}^2$. Show that the inequality fails if the kernel is $|j - k|^{-1}$.

2. Is the form $\sum\limits_{j,\,k\,\geq\,0} b_j c_k (j+k+1)^{-1/2}$ bounded in $\mathbf{l}^2$?

3. For $\eta > 0$ and f in $\mathbf{L}^2(\mathbf{R})$ define

$$T_\eta f(x) = \int\limits_{|t|>\eta} \frac{f(x-t)}{t}\, dt\,.$$

(The integral extends over the line excluding an interval about the origin.) Show that T_η is a bounded operator from $\mathbf{L}^2(\mathbf{R})$ into itself, and for each f, $T_\eta f$ converges to a limit Tf; and T is a bounded operator in $\mathbf{L}^2(\mathbf{R})$. [This is the *conjugacy operator*, formally convolution with the kernel $1/t$. Problem 1 treated a discrete analogue of this operator.]

4. Show that π is the best constant in the Double Series Theorem. [Multiplication by ϕ in $\mathbf{L}^2(\mathbf{T})$ is an operator with bound π. For large n, $(\phi f \chi^n)$ has almost the same norm as its projection on $\mathbf{H}^2(\mathbf{T})$.]

6. Hardy spaces on the line

Hardy spaces on the line, or on the upper halfplane, are more complicated than on the circle or the disk, because there are two natural definitions and they lead to different spaces. Some facts about the circle generalize to one space, others to both.

The conformal map $w = i(1-z)/(1+z)$ carries the unit disk onto the upper halfplane, with 0 taken to i and -1 to ∞. Bounded analytic functions on the disk are carried to the same class on the halfplane. The boundary functions on the line constitute $\mathbf{H}^\infty(\mathbf{R})$, and these are exactly the functions obtained from $\mathbf{H}^\infty(\mathbf{T})$ by the conformal map. There is no ambiguity here. But for finite p the map carries $\mathbf{H}^p(\mathbf{T})$ to a space that we shall call $\mathbf{H}^p_\nu$, which is less easy to describe intrinsically, and is not even part of $\mathbf{L}^p(\mathbf{R})$

(because it contains constant functions, among others). $\mathbf{H}_{\nu}^{p}$ is one kind of Hardy space on the line.

The second kind, called $\mathbf{H}^{p}(\mathbf{R})$, consists of the functions in $\mathbf{L}^{p}(\mathbf{R})$ whose Fourier transforms vanish on the negative real axis. This makes sense immediately for $p = 1$; the Plancherel theorem gives it meaning for $p = 2$; for other values of p the Fourier transform has to be extended in some way. Or else the function-theoretic description of $\mathbf{H}^{p}(\mathbf{T})$ on the disk can be paraphrased on the halfplane; this method will define $\mathbf{H}^{p}(\mathbf{R})$ for all p.

Translations form a group of isometric operators in each space $\mathbf{H}^{p}(\mathbf{R})$. This group structure leads to analogues, with similar proofs, of many theorems about $\mathbf{H}^{p}(\mathbf{T})$. But translation is not isometric in $\mathbf{H}_{\nu}^{p}$ for finite p, and the group structure of $\mathbf{R}$ is not useful in studying these spaces.

Nevertheless the spaces $\mathbf{H}_{\nu}^{p}$ and $\mathbf{H}^{p}(\mathbf{R})$ are related in a simple way: f is in $\mathbf{H}_{\nu}^{p}$ if and only if $f(x)/(1 - ix)^{2/p}$ belongs to $\mathbf{H}^{p}(\mathbf{R})$, and this correspondence is an isometry of the spaces aside from a constant factor. This is the result we are aiming for; it is not obvious. First we have to define the spaces carefully, and then there are function-theoretic difficulties to be met. Once this theorem is proved it is simple to pass information back and forth between the circle and the line.

In Section 2 of Chapter 2 we met the Poisson kernel of the upper halfplane:

$$(6.1) \qquad P_{u}(x) = \frac{u}{\pi (u^{2}+x^{2})} \qquad (u > 0).$$

Note that $P_{u}(x)$ is the real part of $i/\pi(x + iu)$, and is therefore harmonic. We showed that convolution of functions in $\mathbf{L}^{p}(\mathbf{R})$ with this kernel generates harmonic functions in the halfplane. This fact (Problem 1 below) was implicit in the analysis: *for f in* $\mathbf{L}^p(\mathbf{R})$, *the harmonic extension* $F(x + iu)$ is bounded in every interior halfplane $u \geq u_0 > 0$.

Here is the function-theoretic definition of the intrinsic Hardy spaces of the line: $\mathbf{H}^p(\mathbf{R})$ *is the subspace of* $\mathbf{L}^p(\mathbf{R})$ *consisting of all f whose harmonic extensions are analytic in the upper halfplane.* By Problem 1 below, if a sequence (f_n) in $\mathbf{H}^p(\mathbf{R})$ converges in the norm of $\mathbf{L}^p(\mathbf{R})$ to f, then the analytic extensions F_n converge to F uniformly on compact subsets of the halfplane. Therefore F is also analytic, and this shows that $\mathbf{H}^p(\mathbf{R})$ is a closed subspace of $\mathbf{L}^p(\mathbf{R})$. For $p = 1$ we should include the analytic functions that are extensions of finite *measures* on the line; but a version of the theorem of F. and M. Riesz will say that all such measures are absolutely continuous.

The connection between this function-theoretic definition and the Fourier transform is, for $p = 2$, the subject of this classical theorem.

Theorem of Paley and Wiener. *A function f in* $\mathbf{L}^2(\mathbf{R})$ *is in* $\mathbf{H}^2(\mathbf{R})$ *if and only if* $\hat{f}$ *vanishes a.e. on the negative real axis.*

Let f belong to $\mathbf{L}^2(\mathbf{R})$ and have Plancherel transform supported on the positive real axis. We are to show that the harmonic extension $P_u * f$ is analytic. The transform of $P_u * f$ is $(\exp -u|t|)\hat{f}(t)$, supported on the positive real axis. Since the convolution is a smooth function, the inversion theorem shows that

$$(6.2) \qquad P_u * f(x) = \frac{1}{2\pi} \int\limits_0^\infty \hat{f}(t)\, e^{it(x+iu)}\, dt.$$

This function is indeed an analytic function of $x + iu$ in the halfplane.

In the opposite direction, suppose that f is in $\mathbf{H}^2(\mathbf{R})$. Write $\hat{f} = \hat{g} + \hat{h}$, where $\hat{g}$ and $\hat{h}$ are in $\mathbf{L}^2(\mathbf{R})$ and are supported, respectively, on the positive and negative real axis. By what has just been proved, $\hat{g}$ is the transform of a function g in $\mathbf{H}^2(\mathbf{R})$. Thus $h = f - g$ is in $\mathbf{H}^2(\mathbf{R})$ and has transform supported on the *negative* real axis. We must show that $h = 0$.

The transform of $\bar{h}$ is supported on the positive real axis, so $\bar{h}$ is in $\mathbf{H}^2(\mathbf{R})$. Therefore both $\Re h$ and $\Im h$ are in $\mathbf{H}^2(\mathbf{R})$. Their analytic extensions are real, because the Poisson kernel is real. Hence these extensions are constant, and it follows that $\Re h$ and $\Im h$ are constant. These functions are in $\mathbf{L}^2(\mathbf{R})$, so they must be 0. This completes the proof.

Corollary. *For f in $\mathbf{L}^1(\mathbf{R})$, f is in $\mathbf{H}^1(\mathbf{R})$ if and only if $\hat{f}$ vanishes on the negative real axis.*

For positive u, $P_u * f$ has transform $(\exp - u|t|)\hat{f}(t)$, a square-summable function. By the theorem, $P_u * f$ is analytic in each half-plane $u \geq u_0 > 0$ if and only if $\hat{f}$ is supported on the positive real axis, and this proves the theorem.

Theorem of F. and M. Riesz. *Let μ be a measure on the line whose Fourier-Stieltjes transform is supported on the positive real axis. Then μ is absolutely continuous.*

We reduce this result to the theorem on the circle. For any positive number r, define the measure μ_r on the interval $[0, r)$ by setting

$$(6.3) \qquad \mu_r(E) = \sum_{-\infty}^{\infty} \mu(E + kr)$$

for each Borel set E in the interval. The Fourier-Stieltjes coefficients of this measure are

$$(6.4) \qquad a_n(\mu_r) = \int\limits_0^r e^{-2\pi nix/r} \, d\mu_r(x) = \hat{\mu}(2\pi n/r).$$

Hence $a_n(\mu_r) = 0$ for $n \leq 0$; it follows that μ_r is absolutely continuous, for every positive r.

If μ had a nonnull singular part, then by translating μ and choosing r large enough most of the mass of this singular part would be contained in $[0, r)$, so that μ_r would also have a nonnull singular part. This is impossible, and the theorem is proved.

Problem 2 below asks for a translation of this result to a function-theoretic statement.

We come to our main theorem. Let f be a function on $\mathbf{T}$, and F the function on $\mathbf{R}$ defined by

$$(6.5) \qquad F(x) = f(e^{i\theta}) \quad \text{where} \quad x = \frac{i(1 - e^{i\theta})}{1 + e^{i\theta}}.$$

The infinitesimals of x and θ satisfy

$$(6.6) \qquad d\theta = \frac{2\,dx}{1 + x^2}.$$

Hence

$$(6.7) \qquad 2\int\limits_{-\infty}^{\infty} |F(x)|^p \frac{dx}{1 + x^2} = \int\limits_0^{2\pi} |f(e^{i\theta})|^p \, d\theta.$$

This leads to the relation

$$(6.8) \qquad ||F||_{p,\nu} = \pi^{1/p} ||f||_p,$$

where on the left the norm refers to the space $\mathbf{L}^p$ based on the measure $d\nu(x) = (1 + x^2)^{-1}dx$, which we call $\mathbf{L}_\nu^p$, and on the right the norm is in $\mathbf{L}^p(\mathbf{T})$. Thus the conformal map of the circle onto the line carries $\mathbf{L}^p(\mathbf{T})$ isometrically (aside from a factor) onto $\mathbf{L}_\nu^p$.

Theorem 21. *For* $1 \le p < \infty$ *the following subspaces of* $\mathbf{L}_\nu^p$ *are the same:*

(a) *the image of* $\mathbf{H}^p(\mathbf{T})$ *under the conformal map*

(b) *the closed span in* $\mathbf{L}_\nu^p$ *of all exponentials* $\exp iux$ *with* $u \ge 0$

(c) *all functions* $(1 - ix)^{2/p}g(x)$, *where* g *is in* $\mathbf{H}^p(\mathbf{R})$.

This subspace of $\mathbf{L}_\nu^p$ is called $\mathbf{H}_\nu^p$.

Define $\mathbf{H}_\nu^p$ to be the set (a), a closed subspace of $\mathbf{L}_\nu^p$. Since the exponentials of (b) are bounded analytic functions in the upper halfplane they belong to $\mathbf{H}_\nu^p$, so the set (b) is contained in $\mathbf{H}_\nu^p$. We shall show that they span a dense subspace, and this will prove the equivalence of (a) and (b).

The functions χ^n $(n = 0, 1, \dots)$ span $\mathbf{H}^p(\mathbf{T})$. Their images under the conformal map are $(i - x)^n/(i + x)^n$, which therefore span $\mathbf{H}_\nu^p$. In order to prove that the exponentials of (b) are dense in $\mathbf{H}_\nu^p$ it will suffice to show that if f is any function in $\mathbf{L}_\nu^q$ (where q is the exponent conjugate to p) such that

$$(6.9) \qquad \int_{-\infty}^{\infty} f(x)(1 + x^2)^{-1}e^{iux}\,dx = 0 \qquad \text{(all } u \ge 0)$$

then also

$$(6.10) \qquad \int_{-\infty}^{\infty} f(x)(1 + x^2)^{-1}[(x - i)/(x + i)]^n\,dx = 0 \qquad (n = 0, 1, \dots).$$

For $n = 0$, (6.10) is immediate. Write $(x - i)/(x + i) =$

$1 - 2i/(x+i)$. The binomial theorem shows that it will be enough to prove that

$$(6.11) \qquad \int_{-\infty}^{\infty} f(x)(1 + x^2)^{-1}(x + i)^{-n} \, dx = 0 \qquad (n = 1, 2, \dots).$$

For each positive n there is a measure μ_n carried on the positive real axis such that

$$(6.12) \qquad (x + i)^{-n} = \int_{0}^{\infty} e^{itx} \, d\mu_n(t).$$

For $n = 1$, this measure is $-ie^{-t} dt$ on the positive axis, null on the negative axis; and μ_n is the n-fold convolution of μ_1 for $n = 2, 3, \dots$. Thus (6.11) is the same as

$$(6.13) \qquad \int_{-\infty}^{\infty} f(x)(1 + x^2)^{-1} \int_{0}^{\infty} e^{itx} \, d\mu_n(t) \, dx = 0.$$

The integrand is summable on the product space. By Fubini's theorem and the hypothesis (6.9), the integral with respect to x is 0 for all positive t. This proves that (a) and (b) are the same.

The set (c) is a closed subspace of $\mathbf{L}_\nu^p$. For non-negative u, $\exp iux/(1 - ix)^{2/p}$ belongs to $\mathbf{L}^p(\mathbf{R})$ and has bounded analytic extension to the upper halfplane. Therefore this function belongs to $\mathbf{H}^p(\mathbf{R})$ (a proof of this point is asked for below). That is, $\exp iux$ is in the set $(1 - ix)^{2/p} \cdot \mathbf{H}^p(\mathbf{R})$. Hence (b) is contained in (c).

We must show conversely that if g is in $\mathbf{H}^p(\mathbf{R})$, then $g(x)(1 - ix)^{2/p}$ is in $\mathbf{H}_\nu^p$. This will follow from the relations

$$(6.14) \qquad \int_{-\infty}^{\infty} g(x)(1 - ix)^{2/p}[(x - i)/(x + i)]^n (1 + x^2)^{-1} \, dx = 0 \qquad (n \geq 1),$$

for these quantities are the Fourier coefficients with negative indices of the function on the circle that corresponds to $g(x)(1 - ix)^{2/p}$ on the line.

Extend the integrand of (6.14) to the upper halfplane:

$$(6.15) \qquad G(w)(1 - iw)^{2/p}[(w - i)/(w + i)]^n(1 + w^2)^{-1},$$

where G is the analytic extension of g. This expression is a constant times

$$(6.16) \qquad G(w)(w + i)^{-2/q}[(w - i)/(w + i)]^{n-1}.$$

The fraction is bounded in the upper halfplane. By the Hölder inequality, this function has bounded means of order 1 on lines $\Im w = u > 0$. Hence its boundary function, which is a multiple of the integrand in (6.14), is in $\mathbf{H}^1(\mathbf{R})$. But such functions have mean value 0, because their Fourier transforms are continuous and vanish on the negative real axis by the corollary to the theorem of Paley and Wiener. Thus (6.14) holds and the proof is finished.

Problems

1. If f is in $\mathbf{L}^p(\mathbf{R})$, p finite, then its harmonic extension satisfies $|F(x + iu)| \leq A_p u^{-1/p}\|f\|_p$ for positive u. Find a value for A_p.

2. Prove the function-theoretic version of the theorem of F. and M. Riesz: if F is analytic in the upper halfplane with means of order 1 on horizontal lines bounded, then F is the Poisson integral of a summable function.

3. If f is in $\mathbf{H}^\infty(\mathbf{R}) \cap \mathbf{L}^p(\mathbf{R})$, then f is in $\mathbf{H}^p(\mathbf{R})$.

4. Say a function f in $\mathbf{H}^p(\mathbf{R})$ is *outer* if the family $(e^{iux}f(x))$, $u \geq 0$, spans a dense subspace of $\mathbf{H}^p(\mathbf{R})$. (If $p = \infty$, the family should span a subspace that is dense in the weak$*$ topology.) Show that $(1 - ix)^{-2/p}$ is outer in $\mathbf{H}^p(\mathbf{R})$ for finite p.

5. A function q on the line is *inner* if q is in $\mathbf{H}^\infty(\mathbf{R})$ and $|q(x)| = 1$ a.e. Show that every function in $\mathbf{H}^1(\mathbf{R})$ is the product of an inner and an outer function.

6. Describe all inner functions on $\mathbf{R}$ that have no zeros in the upper halfplane and are analytic across the real axis.

7. Show that if a function in $\mathbf{H}^1(\mathbf{R})$ is real, then it is constant.

8. Prove this version of the theorem of Paley and Wiener: let $f(z)$ be entire, square-summable on the real axis, and $O(\exp r|u|)$ for some positive r, where $u = \Im z$. Then the Plancherel transform of f is supported on $(-r, r)$.

Chapter 5
Conjugate Functions

1. Conjugate series and functions

The series *conjugate* to a trigonometric series

$$(1.1) \qquad S = \sum_{-\infty}^{\infty} a_n e^{nix}$$

is

$$(1.2) \qquad T = -i \sum_{-\infty}^{\infty} (\operatorname{sg} n) \, a_n e^{nix},$$

where $\operatorname{sg} n = 1$ if $n > 0$, -1 if $n < 0$, and $\operatorname{sg} 0 = 0$. Formally

$$(1.3) \qquad S + iT = a_0 + 2 \sum_{1}^{\infty} a_n e^{nix};$$

such a series is *of analytic type*.

If the series

$$(1.4) \qquad S(z) = \sum_{-\infty}^{\infty} a_n r^{|n|} e^{nix} \qquad (z = e^{ix})$$

converges in the unit disk it defines a harmonic function. The function

$$(1.5) \qquad T(z) = -i \sum_{-\infty}^{\infty} (\operatorname{sg} n) \, a_n r^{|n|} e^{nix}$$

has the property that $S(z) + iT(z)$ is analytic in the disk. Thus S and T are conjugate harmonic functions, and conjugacy for the trigonometric series (1.1) and (1.2) is formally the relation at the boundary of conjugacy for harmonic functions. Note that T is normalized to have $T(0) = 0$.

If (a_n) is square-summable, then S is the Fourier series of a function f in $\mathbf{L}^2(\mathbf{T})$, T is also the Fourier series of a function g in $\mathbf{L}^2(\mathbf{T})$, and $f + ig = h$ is in $\mathbf{H}^2(\mathbf{T})$. If f is real then so is g, and $f = \Re h$, $g = \Im h$. Whether or not f is real, g is said to be *conjugate* to f. Then $-f + a_0$ is conjugate to g. Moreover if $a_0 = 0$, the Parseval relation shows that $||f||_2 = ||g||_2$. The function conjugate to f is written $\tilde{f}$.

The notion of conjugate function is extended to any pair of functions f and g having a Fourier relation, perhaps generalized in some way, to the series S and T. This chapter is devoted to theorems asserting that functions f in various classes have conjugate functions in the same or some other classes.

There are two ways to give meaning to T when S is the Fourier series of f in some space larger than $\mathbf{L}^2(\mathbf{T})$. The first way is to prove an inequality of the form

$$(1.6) \qquad\qquad ||g||' \leq A||f||''$$

for some two norms, which may be the same, and some constant A, when f is, say, a trigonometric polynomial. Then the conjugacy operation is extended by continuity to the whole space of f.

The second method is based on the observation that conjugacy is a multiplier operation: the Fourier coefficients of g are those of f multiplied by $(-i \operatorname{sg} n)$. Formally,

$$(1.7) \qquad\qquad -i \sum_{-\infty}^{\infty} (\operatorname{sg} n)\, e^{nix} = \cot \tfrac{x}{2}.$$

Therefore we expect

$$(1.8) \qquad g(e^{ix}) = \int f(e^{i(x-t)}) \cot \frac{t}{2} \, d\sigma(t).$$

This integral can sometimes be given meaning, and provides another definition for conjugate function.

The theory of conjugate functions proves inequalities of the form (1.6), investigates the singular integral (1.8), and shows that the two definitions are the same when they are both meaningful.

Theorem 22. *A series conjugate to a Fourier-Stieltjes series is Abel summable almost everywhere.*

We already know (Theorem 4 and Problem 4 of Chapter 1, Section 7) that a Fourier-Stieltjes series is Abel summable a.e. The present theorem is interesting because the conjugate series may not be a Fourier-Stieltjes series.

Let μ be a measure on $\mathbf{T}$ whose Fourier-Stieltjes series is S of (1.1). We are to show that the series T is summable by Abel means for a.e. x. μ is a linear combination of positive measures; it will suffice to prove the result for each part separately, so that we may assume that μ is positive. Then $H(z) = S(z) + iT(z)$, the analytic function constructed above, has for real part the positive harmonic function $S(z) = P_r * \mu$, and imaginary part

$$(1.9) \qquad T(z) = -i \sum_{-\infty}^{\infty} (\operatorname{sg} n) \, a_n \, r^{|n|} \, e^{nix}$$

that is to be proved to have a limit a.e. as r increases to 1. Since $\exp(-H)$ is analytic and bounded in the disk, it has radial limits almost everywhere. This implies that $S(z)$ has a radial limit a.e. (we already know this), and that $T(z)$ has a radial limit wherever

$\exp(-H)$ has a *nonzero* limit. Since the boundary values of a bounded analytic function are zero only on a null set (unless the function is null), the result is proved.

Problems

1. Show that the function conjugate to a bounded function need not be bounded. Deduce that the conjugacy operation, defined on trigonometric polynomials, is not bounded for the norm of $L^1(T)$. [Let Q be the operation that takes every trigonometric polynomial to its conjugate. If Q is bounded in the norm of $L^1(T)$, then the dual operator Q^* would exist as a bounded operator in $L^\infty(T)$. Compute the dual operator.]

2. Using the ideas of the last problem, show that if p and q are conjugate exponents, with p finite, and if Q is a bounded operator in $L^p(T)$, then Q is bounded in $L^q(T)$.

3. Show function-theoretically that if f is in $L^2(T)$ and vanishes on an arc J of T, then $f + i\tilde{f}$ can be continued analytically across J. Therefore $\tilde{f}$ is smooth on J.

4. Find the functions conjugate to (a) $f(e^{ix}) = x$ on $(0, 2\pi)$; (b) $g(e^{ix}) = -1$ on $(-\pi, 0)$, $= 1$ on $(0, \pi)$.

2. Theorems of Kolmogorov and Zygmund

The operation of passing from a function to its conjugate is a linear operation in $L^2(T)$. We have seen (Problem 1 of the last section) that the conjugate of a bounded function may not be bounded, and that conjugacy cannot be extended to $L^1(T)$ so as to be a bounded operator in that space. A famous theorem of M. Riesz, to be proved in the next section, states that conjugacy is a bounded operator in $L^p(T)$ for $1 < p < \infty$. At the extremes $p = 1$,

∞, theorems of Kolmogorov and of Zygmund give precise information even though conjugacy is not bounded in these spaces. These latter results are the subject of this section.

Theorem of Kolmogorov. *There is a constant K such that for all f in $\mathbf{L}^2(\mathbf{T})$, and all positive λ,*

(2.1) $|\tilde{f}(e^{ix})| < \lambda$ *except on a set of measure at most* $\dfrac{K\|f\|_1}{\lambda}$.

The statement is interesting for large λ.

The conformal mapping

(2.2) $$w = 1 + \frac{z-\lambda}{z+\lambda}$$

carries the right halfplane to a disk of radius 1 centered at $w = 1$. The crucial point for our proof is that the part of the halfplane consisting of z such that $|z| \geq \lambda$ is mapped to the half of the disk where $\Re w \geq 1$. The verification is straightforward.

First suppose that f is a non-negative trigonometric polynomial with mean value 1. Form $h = f + i\tilde{f}$, and let H be the analytic extension of h to the unit disk. Then $H(0) = 1$, and we have

(2.3) $$1 + \frac{1-\lambda}{1+\lambda} = \int 1 + \frac{h(e^{ix}) - \lambda}{h(e^{ix}) + \lambda}\, d\sigma(x).$$

Let E be the set in $\mathbf{T}$ where $|\tilde{f}(e^{ix})| \geq \lambda$. On this set also $|h(e^{ix})| \geq \lambda$, so there the integrand in (2.3) has real part at least 1. This real part is non-negative everywhere; hence

(2.4) $$\frac{2}{1+\lambda} = 1 + \frac{1-\lambda}{1+\lambda} \geq \sigma(E).$$

This is still true if the trigonometric polynomial f is non-negative but has mean value less than 1.

Let f be any function in $\mathbf{L}^2(\mathbf{T})$ that is non-negative with mean value at most 1. Then the result just established holds for the Fejér means K_n*f, whose conjugate functions are $K_n*\tilde{f}$. As n tends to ∞, $K_n*\tilde{f}$ tends to $\tilde{f}$ a.e., and we deduce the same inequality on the measure of the set where $|\tilde{f}| \geq \lambda$ (Problem 1 below).

If f is merely real in $\mathbf{L}^2(\mathbf{T})$, but still with norm at most 1 as an element of $\mathbf{L}^1(\mathbf{T})$, write $f = f_+ - f_-$ where f_+ and f_- are the positive and negative parts of f. Wherever $|\tilde{f}| \geq \lambda$, obviously either $(f_+)\tilde{} \geq \lambda/2$ or $(f_-)\tilde{} \geq \lambda/2$. Moreover f_+ and f_- have mean value at most 1. Hence $|\tilde{f}| < \lambda$ except on the union of two sets, each of measure at most $8\pi/(\lambda + 2)$.

Finally, if f is complex we break it into real and imaginary parts, and find that

$$(2.5) \quad |\tilde{f}| < \lambda \text{ except on a set of measure at most } K/(\lambda + 1)$$

for a constant K, whenever f belongs to $\mathbf{L}^2(\mathbf{T})$ and $||f||_1 \leq 1$. If f is any nonnull function in $\mathbf{L}^2(\mathbf{T})$, (2.5) applied to $f/||f||_1$ leads to the statement

$$(2.6) \quad |\tilde{f}| < \lambda \text{ except on a set of measure at most } \frac{K||f||_1}{||f||_1 + \lambda} \, .$$

This proves the theorem, with a small improvement for small λ.

The reference to $\mathbf{L}^2(\mathbf{T})$ in the theorem is convenient because functions in that space have conjugate functions, but this result and a limiting process will lead to a definition of conjugate function for elements of $\mathbf{L}^1(\mathbf{T})$.

Corollary. *For any p between* 0 *and* 1 *there is a constant K such that for all f in* $\mathbf{L}^2(\mathbf{T})$

$$(2.7) \qquad \int |\tilde{f}|^p \, d\sigma \leq K \|f\|_1^p.$$

Assume the corollary, and let f be any function in $\mathbf{L}^1(\mathbf{T})$. Choose a sequence (f_n) of square-summable functions tending to f in $\mathbf{L}^1(\mathbf{T})$. By (2.7), $(\tilde{f}_n)$ is a Cauchy sequence in $\mathbf{L}^p(\mathbf{T})$, a complete metric space (although not normed), so that $\tilde{f}_n$ converges in this space to a limit function that we define to be $\tilde{f}$. It is easy to see that $\tilde{f}$ is independent (modulo sets of measure 0) of the approximating sequence.

Fatou's lemma shows that $\tilde{f}$ satisfies (2.7). Thus the conjugacy operation carries $\mathbf{L}^1(\mathbf{T})$ to $\mathbf{L}^p(\mathbf{T})$, for each p with $0 < p < 1$, and (2.7) holds with a constant K that depends on p.

Let us deduce the corollary from (2.6). It will suffice to prove (2.7) when $\|f\|_1 = 1$. Define $F(\lambda)$ to be the normalized measure of the set where $|\tilde{f}| \geq \lambda$ ($\lambda \geq 0$). Thus F decreases on the positive real axis from 1 to 0. By (2.6), $F(\lambda) \leq K/(\lambda+1)$ for some K independent of f. (The function $1 - F$ is the *distribution function* of $|\tilde{f}|$.) The definition of the Lebesgue integral shows that

$$(2.8) \qquad \int |\tilde{f}|^p \, d\sigma = - \int\limits_0^\infty \lambda^p \, dF(\lambda).$$

(F may well have a discontinuity at 0, but the integrand vanishes there, so the integral is well defined.) The formula for integration by parts (see the Appendix) is applicable: (2.8) equals

$$(2.9) \qquad -\lim_{A \to \infty} \left\{ \left[\lambda^p F(\lambda) \right]_0^A - \int_0^A \left(p\lambda^{p-1} \right) F(\lambda) \, d\lambda \right\}.$$

The bracket tends to 0 as λ tends to ∞, and the integral has a finite limit that is independent of f, because $p < 1$. Thus (2.8) is less than a constant times $\|f\|_1$ for all f in $\mathbf{L}^2(\mathbf{T})$. This proves the corollary.

The next theorem has an interesting nonlinear character.

Theorem of Zygmund. *If f is real and bounded, with bound $\pi/2$, then*

$$(2.10) \qquad \int \exp |\tilde{f}| \cos f \, d\sigma \leq 2.$$

Let f be a real trigonometric polynomial. Then

$$(2.11) \qquad 1 \geq \Re \exp \int if \, d\sigma = \Re \exp \int if - \tilde{f} \, d\sigma =$$
$$\Re \int \exp \left(if - \tilde{f} \right) d\sigma = \int e^{-\tilde{f}} \cos f \, d\sigma.$$

Replacing f by $-f$ gives a similar inequality with $\tilde{f}$ in place of $-\tilde{f}$. Adding these inequalities gives (2.10).

Now let f be any real function bounded by $\pi/2$. choose a sequence (f_n) of real trigonometric polynomials tending to f pointwise and with bound $\pi/2$. (The Fejér means of f are such a sequence.) On a subsequence at least, the conjugates $\tilde{f}_n$ converge to $\tilde{f}$ a.e. (because the sequence converges in $\mathbf{L}^2(\mathbf{T})$). The inequality (2.10), valid for each $\tilde{f}_n$, holds in the limit by Fatou's lemma. Part of the conclusion, of course, is that the integrand in (2.10) is summable.

If $\|f\|_\infty$ is strictly smaller than $\pi/2$, then $\cos f$ is bounded from 0. This implies the first of two corollaries.

Corollary. *If f is real and* $||f||_\infty$ *is strictly smaller than* $\pi/2$, *then*

$$(2.12) \qquad \int \exp |\tilde{f}| \, d\sigma \leq K$$

where K depends only on the bound of f.

Corollary. *If f is continuous, then* $\exp t\tilde{f}$ *is summable for every real t.*

We may assume that f is real. Write $f = g + h$ where g is a real trigonometric polynomial and h has uniform bound less than $\pi/2$. Then $\tilde{g}$ is a trigonometric polynomial, so that $\exp \tilde{g}$ is bounded. Since $\exp \tilde{h}$ is summable by the previous corollary, $\exp \tilde{f}$ is summable. The same argument applies to tf for any t.

Problems

1. (h_n) is a sequence of measurable functions on **T** such that $h_n \leq 0$ except on a set of measure at most η for each n. Suppose that h_n tends to h a.e. Show that h has the same property.

2. Let F be analytic in the unit disk and take values in the sector: $|\arg z| \leq \pi/2 - \epsilon$, where ϵ is positive. Show that F is in $\mathbf{H}^1(\mathbf{T})$.

3. Find a function in $\mathbf{L}^1(\mathbf{T})$ whose conjugate is not in $\mathbf{L}^1(\mathbf{T})$.

4. Find a function f with bound exactly $\pi/2$ such that $\exp |\tilde{f}|$ is not summable.

5. (a) k is a positive constant. Show that there is a real summable function f such that $\tilde{f} = kf$ a.e. (b) There is a real

bounded function f such that $|f| + |\tilde{f}| = 1$ a.e. [Use the Riemann mapping theorem.]

3. Theorems of Riesz and Zygmund

Theorem of M. Riesz. *Let $1 < p < \infty$. For every f in $\mathbf{L}^p(\mathbf{T})$, $\tilde{f}$ is in the same space and*

$$(3.1) \qquad \|\tilde{f}\|_p \leq A_p \|f\|_p$$

where A_p depends only on p.

First assume that f is a positive trigonometric polynomial, so that $h = f + i\tilde{f}$ is a trigonometric polynomial taking values in the right halfplane. Then the analytic extension $H(z)$ also takes values in the halfplane (because it is the Poisson integral of h), and we can form the analytic function $H(z)^p$ using the principal value of the power function. This function is analytic in the interior of the disk, and continuous on the closed disk. By Cauchy's theorem and the fact that $\tilde{f}$ has mean value 0,

$$(3.2) \qquad \int h^p \, d\sigma = H(0)^p = \left[\int f + i\tilde{f} \, d\sigma \right]^p = \left[\int f \, d\sigma \right]^p.$$

At this point we need an elementary lemma.

Lemma. *If $1 < p < 2$ there is a number K such that*

$$(3.3) \qquad 1 \leq (1 + K)(K \cos y)^p - K \cos py$$

for all y such that $|y| \leq \pi/2$.

The inequality is equivalent to

$$(3.4) \qquad (1 + K)^{-1}(1 - \cos py) \leq (K\cos y)^p - \cos py.$$

It will suffice to show that the right side is positive and bounded from 0 for the given range of y, if K is large enough. For $y = 0$ the right side is $K^p - 1$. As y increases, $\cos py$ decreases and reaches 0 while the first term on the right is still positive, because $p > 1$. Over this interval the right side is bounded from 0 if K is large. As y increases further to $\pi/2$, $\cos py$ is negative. This proves the conclusion for y in $[0, \pi/2]$, and both sides of (3.4) are even functions.

Multiply (3.3) by $|h|^p$:

$$(3.5) \qquad |h|^p \leq (1 + K)(K|h| \cos y)^p - K|h|^p \cos py.$$

Choose for y the argument of h. Since the values of h lie in the right halfplane, $|y| \leq \pi/2$. Moreover $|h| \cos y = \Re\, h = f$, and $|h|^p \cos py = \Re\, h^p$. Therefore the integral of (3.5) leads to

$$(3.6) \qquad \|h\|_p^p \leq (1 + K)K^p \|f\|_p^p - K\Re \int h^p \, d\sigma.$$

By (3.2), the integral on the right is the positive number $a_0(f)^p$. Hence the inequality (3.6) is still true without the last term. On the other hand $|\tilde{f}| \leq |h|$, so that (3.1) is proved when f is a positive trigonometric polynomial.

The rest of the proof is like the corresponding argument of the last section. If f is square-summable and non-negative, then $K_n * f$ is a positive trigonometric polynomial whose conjugate is $K_n * \tilde{f}$. These functions satisfy (3.1), and $K_n * \tilde{f}$ converges to $\tilde{f}$ at least on a subsequence. Fatou's lemma shows that (3.1) holds for f

itself.

If f is still real in $\mathbf{L}^2(\mathbf{T})$ but not necessarily of one sign, we write $f = f_+ - f_-$ as before, and (3.1) follows with a larger constant; and then the same follows for complex f. Since $\mathbf{L}^2(\mathbf{T})$ is dense in $\mathbf{L}^p(\mathbf{T})$ for finite p, the theorem is proved for $1 < p < 2$.

Problem 2 of Section 1 established that the same must be true in $\mathbf{L}^q(\mathbf{T})$, where q is the exponent conjugate to p. This completes the proof of the theorem.

For $p = 1$ we know that the analogous result is false. The theorem of Kolmogorov and its corollary of the last section provide one substitute. Here is another substitute, which has a partial converse.

Theorem of Zygmund. *If $f \log^+ |f|$ is summable, then $\tilde{f}$ is summable and*

$$(3.7) \qquad \int |\tilde{f}| \, d\sigma \leq A + B \int |f| \log^+ |f| \, d\sigma$$

for some constants A and B that are independent of f. Conversely, if f is summable and non-negative, and $\tilde{f}$ summable, then $f \log^+ |f|$ is summable.

Recall that $\log^+ x = \max(0, \log x)$ for $x \geq 0$.

First assume, again, that f is a positive trigonometric polynomial, whose mean value we call C, and which we assume ≥ 1. Define $h = f + i\tilde{f}$, an analytic trigonometric polynomial with values in the right halfplane. Define $\log h$ by means of the principal branch of the logarithm. Then $\log h$ is analytic and continuous. By Cauchy's theorem,

$$(3.8) \qquad C \log C = \int h \log h \, d\sigma = \int f \log |h| - y\tilde{f} \, d\sigma,$$

where the argument y of h takes values in $(-\pi/2, \pi/2)$. Note that y has the same sign as $\tilde{f}$. By hypothesis the left side of (3.8) is non-negative, so we have

$$(3.9) \qquad \int y\tilde{f}\,d\sigma \;\leq\; \tfrac{1}{2}\int f\log\left(f^2+\tilde{f}^2\right)d\sigma.$$

In order to defeat the $\tilde{f}$ on the right we need another inequality: *for positive a and b*

$$(3.10) \qquad a\log\left(a^2+b^2\right)\leq 2a\log a + b.$$

A proof is asked for below. (This is not the best such inequality, but it is good enough.) If we set $a=f$ and $b=|\tilde{f}|$, the right side of (3.9) is seen to be less than

$$(3.11) \qquad \int f\log f\,d\sigma + \tfrac{1}{2}\int |\tilde{f}|\,d\sigma.$$

Now we assume further that $f\geq e$ everywhere. Let E be the set where $|\tilde{f}|\leq f$. Obviously

$$(3.12) \qquad \int_E |\tilde{f}|\,d\sigma \leq \int f\,d\sigma \leq \int f\log f\,d\sigma.$$

On the complementary set $|y|\geq\pi/4$, and therefore

$$(3.13) \qquad \int_{CE} |\tilde{f}|\,d\sigma \leq \tfrac{4}{\pi}\int y\tilde{f}\,d\sigma \leq \tfrac{4}{\pi}\left\{\int f\log f\,d\sigma + \tfrac{1}{2}\int |\tilde{f}|\,d\sigma\right\}.$$

Add (3.12) and (3.13), and take the term with $\tilde{f}$ to the left side:

$$(3.14) \qquad \left(1 - \tfrac{2}{\pi}\right) \int |\tilde{f}| \, d\sigma \leq \left(\tfrac{4}{\pi} + 1\right) \int f \log f \, d\sigma.$$

The coefficient on the left is positive, so we have shown that

$$(3.15) \qquad \int |\tilde{f}| \, d\sigma \leq B \int f \log f \, d\sigma$$

for all trigonometric polynomials satisfying $f \geq e$.

Keeping the assumption $f \geq e$, suppose now that f is bounded. Then $(K_n * f)(\log K_n * f)$ converges boundedly to $f \log f$, and (3.15) remains true. If $f \log f$ is summable, and still $f \geq e$, we find a sequence of bounded functions f_n increasing to f, each satisfying $f_n \geq e$, and (3.15) is valid in the limit.

Let f be any real function satisfying the hypotheses of the theorem. Define $f_1 = \max(f, e)$, $f_2 = \min(f, -e)$, $f_3 = f - (f_1 + f_2)$, so that $|f_3| \leq e$. Denote by E_j the set where $f = f_j$ for $j = 1, 2$. Then

$$(3.16) \qquad \int |\tilde{f}_j| \, d\sigma \leq B \int_{E_j} |f| \log |f| \, d\sigma + Be \qquad (j = 1, 2)$$

$$\int |\tilde{f}_3| \, d\sigma \leq \|f_3\|_2 \leq e.$$

Adding these inequalities proves (3.7).

Finally, if f is complex the result follows with new constants from the real case.

The second part of the theorem is a consequence of (3.8). Adding a constant to f if necessary (this does not change $\tilde{f}$), we may assume that $f \geq 1$. Then

$$(3.17) \qquad \int f \log f \, d\sigma \leq \int f \log |h| \, d\sigma = C \log C + \int y \tilde{f} \, d\sigma$$

$$\leq C \log C + \tfrac{\pi}{2} \int |\tilde{f}| \, d\sigma,$$

first when f is a trigonometric polynomial, and then by Fatou's lemma for any non-negative summable function f whose conjugate is summable. This completes the proof.

Problems

1. Prove (3.10).

2. Let ν be a probability measure (a positive measure with total mass 1) on Borel sets of the complex plane. Assume that

$$P(0) = \int P(z)\, d\nu(z)$$

for all polynomials P. Write $z = x + iy$. Show that

$$\int |y|^p\, d\nu(z) \le A_p \int |x|^p\, d\nu(z)$$

where A_p depends only on p, $1 < p < \infty$.

3. Find an upper bound for the best constant A_p in Riesz' theorem.

4. Use Zygmund's theorem of this section to show that $\tilde{f}$ is not summable for certain f in $\mathbf{L}^1(\mathbf{T})$.

5. Show that for f in $\mathbf{L}^1(\mathbf{T})$, the summability of $\tilde{f}$ does not imply the summability of $(f_+)^{\tilde{}}$.

4. The conjugate function as a singular integral

Theorem 23. *Let f be continuous and satisfy a uniform Lipschitz condition on $\mathbf{T}$. Then $\tilde{f}$ is continuous, is the sum of its Fourier series at each point, and has the representation*

$$(4.1) \qquad \tilde{f}(e^{ix}) = \lim_{\epsilon \downarrow 0} \int_{|t| > \epsilon} f(e^{i(x-t)}) \cot(t/2)\, d\sigma(t)$$

(The integral is understood over the interval $(-\pi, \pi)$ with the subinterval $(-\epsilon, \epsilon)$ excluded.)

The convergence of the Fourier series of $\tilde{f}$ can be proved in the same way as Theorem 1 (Problem 1 below). However the other assertions of the theorem require a different treatment, and convergence will be proved at the same time.

The conjugate of the Dirichlet kernel is

$$(4.2) \qquad \tilde{D}_n(e^{ix}) = -i\sum_1^n e^{jix} - e^{-jix} = \cot(x/2) - \frac{\cos\left(n+\frac{1}{2}\right)x}{\sin(x/2)}.$$

Since $\tilde{D}_n$ has mean value 0, the partial sums of the Fourier series of $\tilde{f}$ can be written

$$(4.3) \qquad \tilde{D}_n * f(e^{ix}) = \int [f(e^{i(x-t)}) - f(e^{ix})]\,\tilde{D}_n(e^{it})\,d\sigma(t).$$

The function $[f(e^{i(x-t)}) - f(e^{ix})]/\sin(t/2)$ of t is summable. Therefore

$$(4.4) \qquad \tilde{D}_n * f(e^{ix}) = \int [f(e^{i(x-t)}) - f(e^{ix})]\cot(t/2)\,d\sigma(t)$$
$$- \int [f(e^{i(x-t)}) - f(e^{ix})]\,\frac{\cos\left(n+\frac{1}{2}\right)t}{\sin(t/2)}\,d\sigma(t)$$

where both integrands are summable. The last term tends to 0 as n tends to ∞ by Mercer's theorem. Hence the Fourier series of $\tilde{f}$ converges everywhere; at a point of continuity the sum is $\tilde{f}$, because the series is summable to that value, and cannot converge to anything else. Moreover the sum is $\tilde{f}$ almost everywhere, because $\tilde{f}$ is square-summable, so its Fourier series is summable to $\tilde{f}$ a.e.

Since the cotangent is an odd function, the right side of (4.1) can be written

$$(4.5) \qquad \lim_{\epsilon \downarrow 0} \int_{|t|>\epsilon} [f(e^{i(x-t)}) - f(e^{ix})] \cot (t/2) \, d\sigma(t),$$

equal to the first integral in (4.4) because the integrand is summable. Thus (4.1) holds a.e., and at each point of continuity of $\tilde{f}$. If we show that (4.5), or equivalently the first integral of (4.4), represents a continuous function, then all the assertions of the theorem will be proved.

In the expression

$$(4.6) \qquad \int [f(e^{i(x-t)}) - f(e^{ix}) - f(e^{i(y-t)}) + f(e^{iy})] \cot (t/2) \, d\sigma(t)$$

the integrand tends pointwise to 0 as y tends to x, and the uniform Lipschitz condition makes the dominated convergence theorem applicable to show that the integral tends to 0. Hence $\tilde{f}$ is continuous, and the proof is complete.

Theorem 24. *For every summable function* f, (4.1) *holds almost everywhere.*

This proof is delicate. We shall show that the formula holds at each point where two conditions are satisfied: where the indefinite integral of f is smooth in a certain sense, and where also the conjugate Fourier series is summable by Abel means. Both conditions hold almost everywhere.

Set $\epsilon = 1 - r$ as r varies from 0 to 1. Let $K_r(e^{it}) = \cot (t/2)$ on $(-\pi, \pi)$ excluding the subinterval $(-\epsilon, \epsilon)$, where $K_r = 0$. Without loss of generality we may assume that f has mean value 0, so that an indefinite integral F of f is continuous on **T**. We shall prove this result: *at any point* x *where*

$$(4.7) \qquad |F(x+t) + F(x-t) - 2F(x)| = o(t) \qquad as \ t \to 0$$

we have

(4.8) $(\tilde{P}_r - K_r) * f(e^{ix}) \to 0$ *as r increases to* 1.

The condition (4.7) holds wherever F is differentiable, and so almost everywhere. Also $\tilde{P}_r * f$ tends to $\tilde{f}$ a.e.; Theorem 22 of Section 1 gives the convergence, and the limit is $\tilde{f}$ because we have metric convergence in $\mathbf{L}^p(\mathbf{T})$ $(0 < p < 1)$ to $\tilde{f}$, by the results of Section 2. Thus it will suffice to prove (4.8) under the hypothesis (4.7).

To simplify notation take $x = 0$. The kernel $\tilde{P}_r - K_r$ is odd; therefore the statement to be proved can be written

(4.9) $\lim_{r \uparrow 1} \left\{ \int_0^\epsilon \tilde{P}_r(e^{it}) [f(e^{-it}) - f(e^{it})] \, dt \right.$

$\left. + \int_\epsilon^\pi \left(\tilde{P}_r(e^{it}) - \cot t/2 \right) [f(e^{-it}) - f(e^{it})] \, dt \right\} = 0.$

We shall show that each integral tends to 0.

For the first integral an integration by parts gives

(4.10) $[\tilde{P}_r(e^{it}) (-F(-t) - F(t) + 2F(0)]_0^\epsilon$

$+ \int_0^\epsilon [F(t) + F(-t) - 2F(0)] \, \tilde{P}_r'(e^{it}) \, dt.$

The bracket on the left vanishes for $t = 0$. Now

(4.11) $\tilde{P}_r(e^{it}) = \dfrac{2r \sin t}{1 - 2r \cos t + r^2} .$

A calculation shows that $\tilde{P}_r(e^{i(1-r)}) = O(1/(1-r))$ as r increases to 1. From (4.7) it follows that the first term in (4.10) tends to 0 as

r tends to 1.

Next we note that

$$(4.12) \qquad \tilde{P}_r'(e^{it}) = (d/dt) \sum_1^\infty 2r^n \sin nt = 2 \sum_1^\infty nr^n \cos nt.$$

Thus for all t

$$(4.13) \qquad |\tilde{P}_r'(e^{it})| \le 2 \sum_1^\infty nr^n = \frac{2r}{(1-r^2)}.$$

Hence the second term of (4.10) is majorized by

$$(4.14) \qquad 2r(1-r)^{-2} \int_0^{1-r} E(t)\, t\, dt < 2r(1-r)^{-2} E(1-r) \int_0^{1-r} t\, dt,$$

where $E(t)$ decreases to 0 as t tends to 0. The right side obviously tends to 0 as r increases to 1, so we have proved that the first integral in (4.9) tends to 0.

To estimate the second term we need the identity

$$(4.15) \qquad \tilde{P}_r(e^{it}) - \cot t/2 = \frac{-(1-r)^2 \cot t/2}{1 - 2r\cos t + r^2},$$

obtained directly from (4.11). When r is replaced by $1-\epsilon$, the quantity that must be shown to tend to 0 with ϵ becomes

$$(4.16) \qquad \epsilon^2 \int_\epsilon^\pi \cot t/2 \, [(1-\epsilon)\sin^2 t/2 + \epsilon^2/4]^{-1} [f(e^{it}) - f(e^{-it})]\, dt.$$

Once more we integrate by parts, and this time the expressions are complicated. The product term is

$$(4.17) \qquad \epsilon^2 \Big[\cot t/2 \, [(1-\epsilon)\sin^2 t/2 + \epsilon^2/4]^{-1} [F(t) + F(-t) - 2F(0)] \Big]_\epsilon^\pi.$$

At $t = \pi$ the cotangent vanishes. When $t = \epsilon$, the expression tends to 0 with ϵ on account of (4.7). Thus (4.17) has limit 0.

The integral that remains is, apart from sign,

$$(4.18) \quad \epsilon^2 \int_\epsilon^\pi \frac{(1-\epsilon)\,[\frac{1}{2} + \cos^2 t/2 + \frac{\epsilon^2}{8}\csc^2 t/2]}{[(1-\epsilon)\sin^2 t/2 + \frac{\epsilon^2}{4}]^2} \, [F(t) + F(-t) - 2F(0)]\, dt.$$

The part of the integral from, say, $1/2$ to π is obviously bounded as ϵ tends to 0, so we can integrate merely from ϵ to $1/2$. On this interval the kernel (which is positive) is dominated by

$$(4.19) \qquad\qquad \frac{k_1 + k_2\epsilon^2/t^2}{k_3 t^2 + \epsilon^2/4}$$

for some constants k_j, and F is bounded (that is all we need). Thus it will be enough to show that

$$(4.20) \qquad\qquad \epsilon^2 \int_\epsilon^{1/2} \frac{k_1 + k_2\epsilon^2/t^2}{k_3 t^2 + \epsilon^2/4} \, dt$$

tends to 0 with ϵ.

Change variables by setting $t = u\epsilon$:

$$(4.21) \qquad\qquad \epsilon \int_1^{1/2\epsilon} \frac{k_1 + k_2/u^2}{k_3 u^2 + \frac{1}{4}} \, du.$$

The integral is bounded, so the expression tends to 0 with ϵ. This completes the proof.

Problems

1. Show, as in the proof of Theorem 1, that if f satisfies a Lipschitz condition at a point, then the conjugate of the Fourier series of f converges at this point.

2. Investigate the convergence at 0 of the conjugate of the Fourier series of the function equal to 0 on $(-\pi, 0)$ and 1 on $(0, \pi)$. Discuss the limit (4.1) for this function. Find the conjugate function.

3. Show that $|a_n(K_r)|$ (where K_r is the function defined in this section) is bounded in n and r, $0 < r < 1$, and tends to $-i(\operatorname{sg} n)$ as r inreases to 1. Deduce that $K_r * f$ converges to $\tilde{f}$ in $\mathbf{L}^2(\mathbf{T})$ for each f in the space.

5. The Hilbert transform

For f in $\mathbf{L}^2(\mathbf{R})$, the conjugate function $\tilde{f}$ is defined by the formula

(5.1) $$\hat{\tilde{f}}(x) = -i(\operatorname{sg} x)\hat{f}(x).$$

$(\operatorname{sg} x = 1$ for $x > 0$, $= -1$ for $x < 0$.) Then $\tilde{f}$ is also called the *Hilbert transform* of f. Conjugacy is a multiplier operation, so it commutes with translation, and we expect it to be formally a convolution with some kernel. On the circle the kernel was $\cot t/2$. Here the kernel is $1/\pi t$. The informal justification for this statement is the calculation

(5.2) $$\lim_{\epsilon \downarrow 0} \int_{|t| > \epsilon} t^{-1} e^{-itx}\, dt = -2i(\operatorname{sg} x) \lim_{\epsilon \downarrow 0,\, A \to \infty} \int_{\epsilon}^{A} (\sin t)/t\, dt$$

$$= -\pi i \operatorname{sg} x.$$

(The limit of the second integral was evaluated in Problem 6 of Chapter 2, Section 2.)

Let K_ϵ be the kernel equal to $1/\pi t$ for $|t| > \epsilon$, and to 0 for

$|t| < \epsilon$. For each positive ϵ this function is square-summable. Its Fourier transform is given by the first equality of (5.2):

$$(5.3) \qquad \pi \hat{K}_\epsilon(x) = -2i(\operatorname{sg} x) \int\limits_{|x|\epsilon}^{\infty} (\sin t)/t \, dt.$$

This function is uniformly bounded in x and positive ϵ. Therefore the operation that takes f to $K_\epsilon * f$ is continuous in $\mathbf{L}^2(\mathbf{R})$, and the bounds of these operators lie under a fixed number. As ϵ tends to 0, $\hat{K}_\epsilon$ tends boundedly to $-i(\operatorname{sg} x)$, and so $K_\epsilon * f$ tends in norm to the function $\tilde{f}$ defined by (5.1). (Compare Problem 3 of the last section.) Thus we have proved

Theorem 25. *For each f in $\mathbf{L}^2(\mathbf{R})$, $K_\epsilon * f$ tends in norm to $\tilde{f}$ as ϵ decreases to* 0.

From the definition (5.1), the Fourier transform of $f + i\tilde{f}$ equals $2\hat{f}$ on the positive real axis, and 0 on the negative axis. Thus $f + i\tilde{f}$ is in $\mathbf{H}^2(\mathbf{R})$, and this justifies our definition.

There is an obscure difference between the line and the circle that is sometimes a source of difficulty. On the circle, the conjugate function is normalized to have mean value 0, and this normalization is contained in the representation of the conjugate function as a convolution with the kernel $\cot t/2$. On the line, square-summable functions do not in general have a mean value (because they are not summable), and yet the conjugate function is certainly normalized in some fashion by (5.1). Of course it is normalized by the property that it belongs to $\mathbf{L}^2(\mathbf{R})$! Also, the analytic extension $F(z)$ of $f + i\tilde{f}$ tends to 0 as z tends to ∞ along vertical lines.

This normalization commutes with translation. The normalization on the circle also commutes with translation. However, if we map the disk on the upper halfplane in the usual way, the harmonic function on the disk that vanishes at the origin is mapped to a function on the halfplane that vanishes at i, not at ∞, and this normalization does not commute with translation. Thus (5.1) is not the conformal map of the definition of conjugacy on the circle. A different singular integral, not a convolution, would define conjugacy on the line with the other normalization.

The theorems of Kolmogorov, M. Riesz and Zygmund proved earlier for the circle group have analogues on the line, which we shall not prove.

Problems

1. Calculate the Hilbert transform of the characteristic function of an interval $[a, b]$.

2. Prove that the harmonic extension of any function in $\mathbf{L}^2(\mathbf{R})$ to the upper halfplane tends to 0 along vertical lines.

3. Find an integral representation for the conjugate function normalized to vanish at i.

6. Maximal functions

Let f belong to $\mathbf{L}^1(\mathbf{R})$. Define the two *maximal functions*

$$
f^*(e^{ix}) = \sup_{0<t<\pi} (2t)^{-1} \int_{x-t}^{x+t} |f(e^{iu})|\, du,
$$
(6.1)
$$
f_*(e^{ix}) = \sup_{0<r<1} |f(re^{ix})|.
$$

These functions are defined everywhere if we allow the value ∞, and they depend only on the modulus of f. Each is the supremum of certain averages of $|f|$, and obviously each is greater than or equal to $|f(e^{ix})|$ almost everywhere.

A discovery of Hardy and Littlewood, which has proved very important in the theory of Fourier series and in other parts of analysis, is that f^* and f_* on the average are not much bigger than $|f|$ itself. In form the theorems resemble those proved earlier about the conjugate function. The results will be proved for f^*; those for f_* will follow easily.

Theorem 26. *For positive* λ,

$$(6.2) \qquad f^* \leq \lambda \text{ except on a set of measure at most } 4\,\|f\|_1/\lambda.$$

Let m denote Lebesgue measure on $\mathbf{T}$.

The proof uses the

Vitali Covering Theorem. *Let* $\mathfrak{F}$ *be any non-empty family of intervals on* $\mathbf{T}$, *whose union we call* J. *There is a sequence of pairwise disjoint intervals* (I_n) *of the family such that*

$$(6.3) \qquad\qquad m(\bigcup_1^\infty I_n) > \tfrac{1}{4} m(J).$$

Find an interval I_1 of $\mathfrak{F} = \mathfrak{F}_1$ whose length is at least $\tfrac{3}{4}$ of $\sup_{I \in \mathfrak{F}} m(I)$. Let $\mathfrak{F}_2$ be the family of intervals of $\mathfrak{F}$ that are disjoint from I_1. Find I_2 in $\mathfrak{F}_2$ with length greater than $\tfrac{3}{4}$ of $\sup_{I \in \mathfrak{F}_2} m(I)$. Then define $\mathfrak{F}_3$ as the family of intervals of $\mathfrak{F}$ that do not intersect I_1 or I_2. Continuing in this way we obtain a disjoint sequence (I_n) of members of $\mathfrak{F}$, and we shall show that it satisfies (6.3).

Obviously $m(I_n)$ tends to 0. Thus for large n the intervals of $\mathfrak{F}_n$ all have small length; therefore no interval is in all the $\mathfrak{F}_n$.

For each n, let I'_n be the interval with the same center as

I_n, and four times the length. For each I in $\mathfrak{F}$ there is a least positive integer n (at least 2) such that I is not in $\mathfrak{F}_n$. This means that I intersects I_{n-1}. Since I is in $\mathfrak{F}_{n-1}$, its length is at most $\frac{4}{3}m(I_{n-1})$. Hence I is contained in I'_{n-1}. That is, every interval of $\mathfrak{F}$ is contained in the union of the I'_n, and this proves the covering theorem.

Take f summable and non-negative. For fixed positive λ, at each x such that $f^*(e^{ix}) > \lambda$ let I_x be an interval centered at e^{ix} for which

$$(6.4) \qquad \int_{I_x} f(e^{iu})\,du > \lambda m(I_x).$$

This is a family to which we apply the covering theorem. The measure of the set where $f^* > \lambda$ does not exceed

$$(6.5) \qquad m(J) \le 4m(\bigcup I_n) \le \frac{4}{\lambda}\sum \int_{I_n} f(e^{iu})\,du \le \frac{4}{\lambda}\int f\,dm.$$

This proves the theorem.

Let F be the distribution function of the non-negative summable function f. That is, $F(u)$ is the Lebesgue measure of the set where $f \le u$. Theorem 26 implies this statement about F: *for positive λ,*

$$(6.6)\quad f^* \le 2\lambda \quad \textit{except on a set of measure at most } 2\pi\frac{4}{\lambda}\int_{\lambda+0}^{\infty} u\,dF(u).$$

For set $g = f$ on the set where $f \le \lambda$, $= 0$ elsewhere; and $h = f - g$. We have $f^* \le g^* + h^* \le \lambda + h^*$. Thus $f^* > 2\lambda$ at most on the set where $h^* > \lambda$, whose measure is not greater than

$$(6.7) \qquad \frac{4}{\lambda} \int h\, dm \;=\; 2\pi \frac{4}{\lambda} \int\limits_{\lambda+0}^{\infty} u\, dF(u).$$

Theorem 26 is analogous to the theorem of Kolmogorov about the conjugate function, proved in Section 2. Now we prove analogues of the theorems of M. Riesz and Zygmund.

Theorem 27. *If f is in $\mathbf{L}^p(\mathbf{T})$ with $1 < p < \infty$, then f^* is in $\mathbf{L}^p(\mathbf{T})$. If $f\log^+ |f|$ is summable, then f^* is summable.*

The corresponding theorem about $p = \infty$ is trivial, because if f is bounded then f^* is bounded with the same bound.

The function-theoretic method used to prove the theorems about conjugate functions is not available now. The information is hidden in Theorem 26, however. This method can also be used to prove the theorems about conjugate functions, once Kolmogorov's theorem of Section 2 has been established.

Take f summable and non-negative, with distribution function F. Let G be the distribution function of f^*. For positive λ, (6.6) gives

$$(6.8) \qquad 1 - G(2\lambda) \;\leq\; \frac{4}{\lambda} \int\limits_{\lambda+0}^{\infty} u\, dF(u) \;\leq\; \frac{4}{\lambda^p} \int\limits_{\lambda+0}^{\infty} u^p\, dF(u).$$

(The second inequality holds because $(u/\lambda)^p$ increases with p for $u > \lambda$.) If f is in $\mathbf{L}^p(\mathbf{T})$, $1 \leq p < \infty$, the integral on the right is finite and tends to 0 as λ tends to ∞. Therefore

$$(6.9) \qquad \lambda^p(1 - G(2\lambda)) \to 0 \quad \text{as} \quad \lambda \to \infty.$$

Now

$$(6.10) \qquad ||f^*||_p^p = \int_0^\infty \lambda^p \, dG(\lambda) = [\lambda^p (G(\lambda) - 1)]_0^\infty$$
$$+ p \int_0^\infty (1 - G(\lambda))\lambda^{p-1} \, d\lambda.$$

The first term on the right vanishes by (6.9). In the second, the integrand is non-negative, and the first inequality of (6.8) gives

$$(6.11) \qquad ||f^*||_p^p = p2^p \int_0^\infty (1 - G(2\lambda))\lambda^{p-1} \, d\lambda$$
$$\leq p2^{p+2} \int_0^\infty \lambda^{p-2} \int_{\lambda+0}^\infty u \, dF(u) \, d\lambda.$$

Suppose $p > 1$. If we change the order of integration in the last term of (6.11) we find

$$(6.12) \quad p2^p \int_0^\infty \int_0^u \lambda^{p-2} \, d\lambda \, u \, dF(u) = \frac{p2^{p+2}}{p-1} \int_0^\infty u^p \, dF(u) = \frac{p2^{p+2}}{p-1} ||f||_p^p,$$

which proves the first part of the theorem.

For $p = 1$, (6.11) becomes

$$(6.13) \quad ||f^*||_1 = 2 \int_0^\infty (1 - G(2\lambda)) \, d\lambda \leq 2 + 2 \int_1^\infty (1 - G(2\lambda)) \, d\lambda$$
$$< 2 + 8 \int_1^\infty \lambda^{-1} \int_{\lambda+0}^\infty u \, dF(u) \, d\lambda.$$

If we change the order of integration we find the other assertion of the theorem.

Problems

1. Prove the assertion used in the text above: if f and g are non-negative functions, then $(f + g)^* \leq f^* + g^*$.

2. Show that $f_* \leq f^*$. (Then the theorems of this section

follow immediatly for f_*.) [Show that every function P that is positive, even and differentiable on $(-\pi, \pi)$, decreasing on $(0, \pi)$ with $P(\pi) = 0$ can be represented in the form

$$P(x) = \int_0^\pi h(t)\, S(t, x)\, dt$$

where $S(t, x) = 1/2t$ for $|x| < t, = 0$ for $|x| > t$ $(0 < t < \pi)$, and h is a non-negative function whose integral over $(0, \pi)$ equals the integral of P over $(-\pi, \pi)$. Apply this result to $P(x) - P(\pi)$ where P is a member of the Poisson kernel.

7. Rademacher functions; absolute Fourier multipliers

The Rademacher functions (φ_k) $(k = 1, 2, \ldots)$ are an orthonormal system in $\mathbf{L}^2(0, 1)$. φ_1 equals 1 on $(0, 1/2)$ and -1 on $(1/2, 1)$. Continue φ_1 to be periodic with period 1. Then $\varphi_k(t) = \varphi_1(2^{k-1}t)$ for each positive integer k. Equivalently, $\varphi_k(t) = 1$ on the dyadic intervals $(j/2^k, (j+1)/2^k)$ for even j, and $= -1$ on such intervals when j is odd $(j = 1, \ldots, 2^k-1)$. The values of the functions at endpoints of the intervals are unimportant; for definiteness we can say that $\varphi_k(t) = 0$ at each point of discontinuity.

Each real t in $(0, 1)$ not a dyadic rational has a unique expansion

$$(7.1) \qquad\qquad t = \sum_1^\infty \varepsilon_k/2^k, \qquad \varepsilon_k = 0 \text{ or } 1.$$

The ε_k tell where t is situated in the interval. If $\varepsilon_1 = 0$, then t is in $(0, 1/2)$; if 1, t is in $(1/2, 1)$. Similarly, $\varepsilon_2 = 0$ if t is in the first half of that subinterval, and $= 1$ if it is in the second half. And so on for each k.

We verify easily that $\varepsilon_k = (1 - \varphi_k)/2$. Both ε_k and φ_k are functions of t.

It is useful to use the language of probability. An *event* is a measurable subset of the *probability space* $(0, 1)$ (with Lebesgue measure). The *probability* of an event is the Lebesgue measure of the set. For example, the probablity that $\varphi_3 = -1$ is $1/2$, because the set of t for which $\varphi_3(t) = -1$ has measure $1/2$.

A *random variable* is a measurable function on a probability space. Our random variables, such as φ_k and ε_k, will be real.

Lemma 1. *The random variables φ_k are independent.*

The statement means: given any measurable subsets $E_1, \ldots, E_n$ of $\mathbf{R}$, and any distinct positive indices $k_1, \ldots, k_n$, the probability that *all* the events

$$(7.2) \qquad \varphi_{k_j}(t) \text{ belongs to } E_j \qquad (j = 1, \ldots, n)$$

occur equals the product of the probabilities of each one. In this case it suffices to test sets E_j each consisting of one point, 1 or -1. From the definition of the Rademacher functions it is obvious by induction that the set where $\varphi_{k_1} = \alpha_1, \ldots, \varphi_{k_n} = \alpha_n$ (each α_k equal to 1 or -1) has measure exactly $1/2^n$.

Much of what follows is simply a transcription, for this special case, of results that hold for any independent sequence of random variables.

Lemma 2. *If $k_1, \ldots, k_n$ are distinct positive integers, then*

$$(7.3) \qquad \int_0^1 \varphi_{k_1} \cdots \varphi_{k_n} \, dt = 0.$$

Suppose that k_n is the greatest integer of the set. On each interval of constancy of $\varphi_{k_1} \cdots \varphi_{k_{n-1}}$, φ_{k_n} takes the values 1 and −1 each with probability 1/2, so the integral (7.3) vanishes over the interval. Hence the integral over (0, 1) is 0.

The square of a Rademacher function is identically 1 (aside from points of dicontinuity). From this fact and Lemma 2 it follows that *two products of distinct Rademacher functions are orthogonal in* $\mathbf{L}^2(0, 1)$, *unless the products are identical.* The family of all products of distinct Rademacher functions, together with the constant function 1, is called the *Walsh system.*

Obviously the Rademacher system is not complete (that is, the Parseval relation does not hold), but Bessel's inequality is true with the same proof as for the trigonometric system. Also the Riesz-Fischer theorem holds: given any square-summable sequence $(c_1, c_2, \dots)$, the series

$$(7.4) \qquad \sum_1^\infty c_k \varphi_k$$

converges in $\mathbf{L}^2(0, 1)$ to a function f, and

$$(7.5) \qquad c_k = \int_0^1 f(t)\, \varphi_k \, dt \qquad (k = 1, 2, \dots).$$

The main theorem, from which we shall obtain results about trigonometric series, is this.

Theorem 28. *If* (c_k) *is square-summable, then the series* (7.4) *converges almost everywhere to its metric sum* f. *Otherwise the series diverges a.e., and indeed a.e. is not even summable by Abel means.*

The Abel summability method is not particular here; the series is not summable by any ordinary method, but our application will involve the Abel method.

Denote the partial sum of order k of (7.4) by S_k. Since S_k tends to f in the metric of $\mathbf{L}^2(0, 1)$, it converges also in $\mathbf{L}^1(0, 1)$ (by the Schwarz inequality). Let $0 \leq a < b \leq 1$, and fix a positive integer k. We have

$$(7.6) \qquad \int_a^b f - S_k \, dt = \lim_n \int_a^b S_n - S_k \, dt.$$

If (a, b) is a maximal interval of constancy of φ_k, then for $n > k$ the integral on the right vanishes, because the difference involves only $\varphi_{k+1}, \ldots, \varphi_n$. Therefore f and S_k have the same mean value over every such interval.

Let F be an indefinite integral of f, and t a point of $(0, 1)$, not a dyadic rational number, where F' exists. For each positive integer k, let J_k be the interval of constancy of φ_k that contains t, whose length we call $|J_k|$. Then

$$(7.7) \qquad S_k(t) = |J_k|^{-1} \int_{J_k} S_k \, dx = |J_k|^{-1} \int_{J_k} f \, dx,$$

and we recognize the right side as a difference quotient of F. As k tends to ∞, this quotient tends to $F'(t)$. Since F is differentiable almost everywhere, and the derivative is $f(t)$ a.e., the first part of the theorem is proved.

Now suppose that the series converges in a set of positive measure. On a subset E still of positive measure m, the partial sums all satisfy $|S_k| \leq M$ for some number M. Hence

$$(7.8) \qquad I = \int_E (S_{k+p} - S_k)^2 \, dt \leq 4mM^2$$

for all positive integers k, p. Expanding the square gives

$$(7.9) \qquad I = \int_E \left(\sum c_j \varphi_j\right)^2 dt = m \sum c_j^2 + 2 \sum c_r c_s \int_E \varphi_r \varphi_s \, dt$$

where the summations extend over indices satisfying $k+1 \le j \le k+p$; $k+1 \le r < s \le k+p$.

The system $(\varphi_r \varphi_s)$ $(r < s)$ is orthonormal by Lemma 2, and the integral on the right side of (7.9) is the Fourier coefficient of the characteristic function of E. These coefficients d_{rs} form a square-summable sequence by Bessel's inequality. Hence if k is large enough

$$(7.10) \qquad \sum d_{rs}^2 \le m^2/9 \qquad (k+1 \le r < s < \infty).$$

By the Schwarz inequality, the modulus of the last term of (7.9) does not exceed

$$(7.11) \qquad 2\left(\sum c_r^2 c_s^2\right)^{1/2} \frac{m}{3} \le \frac{2m}{3} \sum c_j^2,$$

where j, r, s range over the same intervals as in (7.9) Hence

$$(7.12) \qquad 4mM^2 \ge I \ge m \sum c_j^2 - \frac{2m}{3} \sum c_j^2 = \frac{m}{3} \sum_{k+1}^{k+p} c_j^2$$

for all positive integers p, provided that k is so large that (7.10) holds. This proves that (c_k) is square summable, if (7.4) converges on a set of positive measure, or indeed if merely the partial sums are bounded on a set of positive measure.

Suppose only that the Abel means of (7.4) are bounded on a set E of positive measure. That is, if we replace each c_k by $c_k u^k$, then $|S_k(t, u)| \leq M$ for all k, for all t in E, and for all u, $0 < u < 1$. We follow the same steps, noting that the choice of k to insure the truth of (7.10) does not depend on u. Therefore (7.12) holds with the factor u^{2j} in each term of the sums. As u increases to 1 we find the same result.

The theorem can be paraphrased in probabilistic language: *if the signs in the series*

$$(7.13) \qquad \sum_1^\infty \pm c_k$$

are chosen independently and with equal probabilities, then the series converges almost surely if (c_k) is square-summable, and diverges almost surely otherwise.

The Rademacher system is an interesting example of an orthonormal system that is quite different from the trigonometric system, because its members are independent, but it is actually a useful tool for studying the trigonometric system itself. Here are some results that follow from Theorem 28.

Theorem 29. *The series*

$$(7.14) \qquad \sum_{-\infty}^\infty \pm c_k e^{kix}$$

is a Fourier-Stieltjes series for all choices of signs if and only if (c_k) is square-summable.

If the sequence is square-summable, then for any choice of signs the series is the Fourier series of a function in $\mathbf{L}^2(\mathbf{T})$; this direction is trivial.

Suppose that (7.14) is a Fourier-Stieltjes sequence for each choice of signs. Rewrite the series as

(7.15) $$\sum_{-\infty}^{\infty} \psi_k(t)\, c_k\, e^{kix},$$

where (ψ_k) is a reordering of the Rademacher functions in a doubly infinite sequence, so that every choice of signs in (7.14) corresponds to a number t in $(0, 1)$ that is not a dyadic rational (we exclude sequences of signs that terminate in $+$'s or $-$'s). In Chapter 1 we proved that a Fourier-Stieltjes series is summable a.e. by Abel means (Theorem 4 and Problem 4 of Section 7), so for each t (7.15) is summable a.e. By Fubini's theorem, for a.e. x the series is summable for a.e. t. Theorem 28 now implies that (c_k) is square-summable.

The dual of this result is: *a sequence (w_k) of non-negative numbers is square-summable if and only if*

(7.16) $$\sum_{-\infty}^{\infty} |c_k|\, w_k < \infty$$

whenever (c_k) is the Fourier sequence of a continuous function. An interesting strengthening of this statement was proved by R.E.A.C. Paley.

Paley's Theorem. *Let (w_k) $(k \geq 0)$ be a non-negative sequence such that (7.16) holds (the sum now being over $k \geq 0$) whenever*

(7.17) $$\sum_{0}^{\infty} c_k\, e^{kix}$$

is the Fourier series of a continuous function. Then (w_k) is square-summable.

If (7.17) were the Fourier series of a continuous function whenever

(7.18)
$$\sum_{-\infty}^{\infty} c_k e^{kix}$$

is, then Paley's theorem would be an uninteresting corollary of the statement dual to Theorem 29. But this is not the case; the analogue of the theorem of M. Riesz on conjugate series of class $\mathbf{L}^p(\mathbf{T})$, $1 < p < \infty$, is not true in $\mathbf{C}(\mathbf{T})$. Therefore the hypothesis is now genuinely weaker, and Paley's theorem is not obvious.

Define a linear functional F in the subspace of $\mathbf{C}(\mathbf{T})$ consisting of all functions with analytic Fourier series (7.17) by setting

(7.19)
$$F(g) = \sum_{0}^{\infty} a_k(g)\, w_k.$$

The series converges absolutely by hypothesis. Moreover F is the limit of

(7.20)
$$F_n(g) = \sum_{0}^{n} a_k(g)\, w_k.$$

Each F_n is a continuous functional; by the Banach-Steinhaus theorem, F is continuous.

F has a continuous extension to $\mathbf{C}(\mathbf{T})$ by the Hahn-Banach theorem. This functional, by the theorem of F. Riesz, is realized by integration with a measure μ. Taking exponentials for g in (7.19) shows that $a_k(\mu) = w_k$ $(k \geq 0)$. The same argument can be applied to $(\pm w_k)$ for any choice of signs. That is, the series

(7.21)
$$\sum_{0}^{\infty} \pm w_k e^{kix}$$

is half a Fourier-Stieltjes series, for every choice of signs.

Unfortunately the other half of the Fourier-Stieltjes series depends on the choice of signs, and (7.21) is not itself necessarily a Fourier-Stieltjes series. The argument used to prove Theorem 29 fails at this point.

Theorem 22 of Section 1 provides the needed information: *the series conjugate to a Fourier-Stieltjes series is also Abel summable a.e.* It follows that (7.21) is summable a.e., for each choice of signs, and the proof ends as before by means of Fubini's theorem.

Here is a second important result about the Rademacher system.

Theorem 30. *If (c_k) is square-summable, then the sum f of (7.4) belongs to $\mathbf{L}^q(0,1)$ for every finite q, and indeed $\exp u f^2$ is summable for every positive u.*

As soon as the sum of a Rademacher series is square-summable, it is the next thing to being a bounded function.

We may take the c_k real. As before, let S_k be the partial sum of (7.4). We want to estimate the integral of

$$(7.22)\quad S_k^{2q} = (c_1\varphi_1 + \ldots + c_k\varphi_k)^{2q} = \sum \frac{(2q)!}{r_1! \cdots r_k!}\, c_1^{r_1} \cdots c_k^{r_k} \varphi_1^{r_1} \cdots \varphi_k^{r_k};$$

the summation extends over all sets of non-negative integers $r_1, \ldots, r_k$ with sum $2q$. When a term of this sum is integrated, the result is 0 unless *each r_j is even*. For such terms the product of the Rademacher functions is 1, and we find

$$(7.23)\qquad \int_0^1 S_k^{2q}\, dt = \sum \frac{(2q)!}{(2r_1)! \cdots (2r_k)!}\, c_1^{2r_1} \cdots c_k^{2r_k},$$

where now the r_j are non-negative with sum q.

Also

$$(7.24) \qquad (c_1^2 + \ldots + c_k^2)^2 = \sum \frac{q!}{r_1! \cdots r_k!} c_1^{2r_1} \cdots c_k^{2r_k}$$

where the r_j are non-negative and have sum q. Compare the fractions in (7.23) and (7.24); an elementary estimate gives

$$(7.25) \qquad \frac{(2q)!}{(2r_1)! \cdots (2r_k)!} \div \frac{q!}{r_1! \cdots r_k!} \le q^q$$

for all indices $r_1, \ldots, r_k$. Therefore

$$(7.26) \qquad \|S_k\|_{2q}^{2q} \le q^q \|c\|_2^{2q},$$

and

$$(7.27) \qquad \int_0^1 \exp S_k^2 \, dt = \sum_{q=0}^\infty \frac{1}{q!} \|S_k\|_{2q}^{2q} \le \sum_0^\infty \frac{(sq)^q}{q!}$$

where $s = \|c\|_2^2$. If $s < 1$, the sum on the right is finite.

This inequality is independent of k. As k tends to ∞, S_k converges a.e. to f, and (7.27) remains true for f by Fatou's lemma. If $s \ge 1$, we can remove a finite number of terms from (7.4), which does not affect the summablility of $\exp f^2$, and obtain the same conclusion. Thus the theorem is proved for $u = 1$. But the hypothesis is homogeneous with respect to the sequence (c_k), so the conclusion holds for all u.

Pointwise convergence and summability theorems are always more difficult than the corresponding metric theorems. In classical harmonic analysis, pointwise results were often appealed to in contexts where metric versions would suffice, and indeed be more natural. This has led to a modern tendency to neglect the pointwise theorems. This is a pity, because the deeper pointwise theorems have interesting consequences. It is hoped that the results of this section illustrate the point.

Problems

1. Why are the functionals F_n defined by (7.20) continuous? Why does that imply that F is continuous?

2. Write down an unbounded positive function f on $(0, 1)$ such that $\exp uf^2$ is summable for every positive u.

3. Denote by $\mathbf{L}$ the space of real functions f on $(0, 1)$ that are sums of series (7.4) with square-summable coefficient sequences. Show that the p-norm and the q-norm are equivalent on $\mathbf{L}$ for $1 \le p < q < \infty$.

4. Prove this point, used in the text: if the Abel means of (7.4) converge on a set of positive measure, then they are uniformly bounded on some smaller set of positive measure.

5. Verify (7.25).

Chapter 6

Translation

1. Theorems of Wiener and Beurling; the Titchmarsh convolution theorem

For each real t define the translation operator: $T_t f(x) = f(x-t)$ for all functions f defined on $\mathbf{R}$. Translation operators on $\mathbf{Z}$ and on $\mathbf{T}$ are defined similarly. N. Wiener began the study of closed subspaces of $\mathbf{L}^1(\mathbf{R})$ and of $\mathbf{L}^2(\mathbf{R})$ that are invariant under all translations T_t. A. Beurling studied subspaces that are invariant under translations in just one direction on $\mathbf{R}$ or $\mathbf{Z}$; these subspaces are more complicated and interesting. Translation is carried by the Fourier transform to multiplication by exponentials. Thus much of Chapter 4 was about such subspaces. The first objective of this chapter is to characterize the closed subspaces of $\mathbf{L}^2(\mathbf{R})$ invariant under all translations, or under translations to the right. These results are analogous to theorems of Chapter 4 on the circle.

Theorem 31. *Let* $\mathbf{M}$ *be a closed subspace of* $\mathbf{L}^2(\mathbf{R})$ *invariant under all translations. Then* $\mathbf{M}$ *consists of all functions whose transforms are supported on some measurable subset of* $\mathbf{R}$.

The converse of the theorem is obvious.

Let f be an element of $\mathbf{M}$ and g orthogonal to $\mathbf{M}$. The invariance of $\mathbf{M}$ implies that

$$(1.1) \qquad f * \tilde{g}(t) = \int_{-\infty}^{\infty} f(x+t)\,\overline{g}(x)\,dx$$

vanishes identically. (Recall that $\tilde{g}(x) = \overline{g}(-x)$.) The Fourier transform of the convolution is $\hat{f}\hat{\tilde{g}}$, which must therefore be null. That is, transforms of **M** and of $\mathbf{M}^{\perp}$ have disjoint supports. As in the proof of Theorem 18, Chapter 4, this means that all the functions of **M** have transforms with supports in a set E, and those of $\mathbf{M}^{\perp}$ in the complement of E. Hence **M** contains *all* the functions whose transforms are supported in E, as the theorem requires.

Theorem 32. *Let* **M** *be a closed subspace of* $\mathbf{L}^2(\mathbf{R})$ *that is invariant under* T_t *for* $t > 0$. *Either* **M** *is invariant under all* T_t (*and has the form of Theorem 31*), *or* **M** *consists of the transforms of functions in* $q \cdot \mathbf{H}^2(\mathbf{R})$, *for some unitary function* q.

Let **N** denote the set of inverse Fourier transforms of functions of **M**. Invariance of **M** under a translation T_t is the same as invariance of **N** under multiplication by $\exp itx$. If q is any function of modulus 1, $\mathbf{N} = q \cdot \mathbf{H}^2(\mathbf{R})$ is invariant under multiplications by exponentials with positive frequencies; therefore the space **M** of Fourier transforms of **N** is invariant under T_t for $t > 0$. This is the easy half of the theorem.

The other direction is harder than the proof of Theorem 18 for the circle group. Let **N** be a closed subspace invariant under multiplication by $\exp itx$ for all positive t.

Lemma. **N** *admits multiplication by all functions of* $\mathbf{H}^{\infty}(\mathbf{R})$.

For f in **N** and g in $\mathbf{N}^{\perp}$ we have

$$(1.2) \qquad \int_{-\infty}^{\infty} f(x)\, \overline{g}(x)\, e^{itx}\, dx = 0 \qquad (t > 0).$$

Thus $f\overline{g}$ belongs to $\mathbf{H}^1(\mathbf{R})$. Then $f\overline{g}h$ is in $\mathbf{H}^1(\mathbf{R})$ for any h in $\mathbf{H}^{\infty}(\mathbf{R})$, and necessarily

$$(1.3) \qquad \int_{-\infty}^{\infty} f\overline{g}h\, dx = 0.$$

That is, g is orthogonal to fh. Hence fh is in $\mathbf{N}$, as we were to prove.

The lemma enables us to reduce the proof of the theorem to the circle. Multiplication by $x - i$ carries $\mathbf{N}$ isometrically onto a subspace $\tilde{\mathbf{N}}$ of $\mathbf{L}_\nu^2$ (the Lebesgue space $\mathbf{L}^2(\mathbf{R})$ formed with the measure $d\nu(x) = (1 + x^2)^{-1} dx$). Then $\tilde{\mathbf{N}}$ is also invariant under multiplications from $\mathbf{H}^\infty(\mathbf{R})$. The conformal map from the upper halfplane to the unit disk carries $\tilde{\mathbf{N}}$ to a closed subspace of $\mathbf{L}^2(\mathbf{T})$ that is invariant under multiplications from $\mathbf{H}^\infty(\mathbf{T})$. Theorem 18 of Chapter 4 gives the form of this subspace. Going backwards to the line leads to the conclusion that $\mathbf{N}$ consists of all functions of $\mathbf{L}^2(\mathbf{R})$ supported on some subset of the line, or else is $q \cdot \mathbf{H}^2(\mathbf{R})$ for some unitary function q. This proves the theorem.

Theorem 32 provides an easy proof of an important theorem. Here is a first version; Problem 5 below completes the result.

Titchmarsh Convolution Theorem. *Let f and g belong to $\mathbf{L}^2(\mathbf{R})$ and vanish on the left half-axis. If neither f nor g vanishes throughout any interval $(0, r)$ with $r > 0$, then the same is true of $f*g$.*

If f vanishes on an interval $(-\infty, r)$ and g on $(-\infty, s)$, then from the definition of convolution $f*g$ vanishes on $(-\infty, r+s)$. The theorem states that if r and s are exact, then so is $r + s$.

We assume that f and g are not identically 0 on any open interval containing 0, but that $f*g = 0$ on $(0, r)$, $r > 0$. Fix f, and let $\mathbf{M}$ be the set of all h in $\mathbf{L}^2(\mathbf{R})$ that vanish on the left half-axis and such that $f*h = 0$ on $(0, r)$. $\mathbf{M}$ is a closed subspace containing g, and also all functions of $\mathbf{L}^2(\mathbf{R})$ vanishing on $(-\infty, r)$; and $\mathbf{M}$ is invariant under translations to the right. Obviously $\mathbf{M}$ is not invariant under all left-translations. Hence $\mathbf{M}$ is the set of Fourier transforms of $q \cdot \mathbf{H}^2(\mathbf{R})$ for some unitary function q, and q is inner because $\mathbf{M}$ is contained in the set of transforms of $\mathbf{H}^2(\mathbf{R})$. Thus we have

$$(1.4) \qquad e^{irx} \cdot \mathbf{H}^2(\mathbf{R}) \subset q \cdot \mathbf{H}^2(\mathbf{R}) \subset \mathbf{H}^2(\mathbf{R}).$$

It follows that both q and $q^{-1} e^{irx}$ are inner functions. If q contained any Blaschke product, or any singular factor generated by a measure on the line, this would be false. The only other possibility is that q is itself an exponential: $q(x) = e^{isx}$ with $0 \le s \le r$.

Hence all the functions of $\mathbf{M}$ vanish on $(0, s)$. Since g belongs to $\mathbf{M}$, s must be 0. In other words, $f * h$ vanishes on $(0, r)$ for all h supported on the right half-axis. This would imply that f vanishes on $(0, r)$, which is not the case, and the proof is finished.

Problems

1. Show that the closed span of translates of f in $\mathbf{L}^2(\mathbf{R})$ is all of $\mathbf{L}^2(\mathbf{R})$ if and only if $\hat{f} \ne 0$ a.e.

2. Show in detail why (1.4) implies that q has the form $\exp isx$ for some s, $0 \le s \le r$.

3. Let f, g be square-summable, and I, J the smallest intervals supporting f, g, respectively. Assume that I and J are finite intervals. Show that the smallest interval supporting $f * g$ is $I + J$, the interval consisting of all $r + s$ with r in I and s in J. Is the result true if I or J is an infinite interval?

4. Characterize the closed subspaces of $\mathbf{L}^2(\mathbf{T})$ invariant under translations. [There is no corresponding one-sided result on $\mathbf{T}$.]

5. If f and g are locally summmable (that is, summable on

each finite interval) and vanish on the left half-axis, but not throughout any interval $(0, r)$ with $r > 0$, then $f * g$ is defined a.e., and does not vanish identically on any such interval. [After convolution with an appropriate approximate identity, f and g will be locally square-summable and can be modified to be square-summable.]

2. The Tauberian theorem

Much of the deepest analysis in our subject concerns the Banach algebras l^1 and $L^1(R)$. A central result is the Tauberian theorem of N. Wiener, which will be proved in this section.

Let f be a continuous function on T whose Fourier series converges *absolutely*. That is, f is the Fourier transform of an element of l^1. If f is different from 0 everywhere, than $1/f$ is at least continuous. Wiener's first result is that *the Fourier series of $1/f$ is also absolutely convergent*. P. Lévy completed this result to state: *if H is a function analytic on a domain of the complex plane containing the values of f, then $H(f)$ has absolutely convergent Fourier series*. Later the Wiener-Lévy theorem was generalized to Banach algebras, and it is an important tool of modern analysis.

Wiener's theorem does not have an obvious extension to $L^1(R)$ because the inverse of any Fourier transform is unbounded, and cannot be a transform. However, this weaker statement holds: *if F is a Fourier transform not vanishing on an interval $[a, b]$, then there is a transform G such that $FG = 1$ on $[a, b]$*. This is almost

Wiener's Tauberian Theorem. *If f is in $L^1(R)$ and its Fourier transform never vanishes, then translates of f span a dense subspace of $L^1(R)$.*

The name of the theorem will be explained later.

The converse of the theorem is trivial. If $\hat{f}(y) = 0$, then every translate of f has transform vanishing at y, the same is true for linear combinations of translates, and of a limit in norm of such linear combinations. Thus the translates of f cannot span a dense subspace of $\mathbf{L}^1(\mathbf{R})$.

The proof proceeds in a series of steps, several of interest as independent results.

(a) *Closed ideals of* $\mathbf{L}^1(\mathbf{R})$ *are identical with closed subspaces invariant under translations.*

The proof is straightforward and is asked for below.

(b) *The set of functions in* $\mathbf{L}^1(\mathbf{R})$ *whose transforms are compactly supported is dense in the space.*

This was Problem 2 of Chapter 2, Section 2. Let (e_n) be an approximate identity consisting of functions whose transforms are compactly supported, for example the Fejér kernel. Then e_n*f has compactly supported transform, and tends to f, for each f.

(c) *Let* f *belong to* $\mathbf{L}^1(\mathbf{R})$ *and suppose that* $\hat{f}$ *does not vanish on the interval* $[a, b]$. *Then there is a function* g *in* $\mathbf{L}^1(\mathbf{R})$ *such that* $\hat{f}\hat{g} = 1$ *on that interval.*

This will prove the Tauberian theorem. For let f be an element whose transform never vanishes. Denote the smallest closed ideal containing f by $\mathbf{M}_f$. Let h be any function in the space whose transform has compact support, and g a function such that $\hat{f}\hat{g} = 1$ on an interval containing the support of $\hat{h}$. Then $\hat{h} = \hat{f}\hat{g}\hat{h}$, so that h belongs to $\mathbf{M}_f$. Such functions h are dense in $\mathbf{L}^1(\mathbf{R})$ by (b), and this proves the theorem.

Thus we must prove (c).

(d) A *triangular function* is a function on $\mathbf{R}$ whose graph is

a triangle with base on the horizontal axis, and having equal
sides. Such a function is a translate of the Fourier transform of a
Fejér kernel. The *trapezoidal function* that vanishes outside
$(-2, 2)$, equals 1 on $(-1, 1)$, and is linear on the intervals $(-2, -1)$,
$(1, 2)$ is a difference of triangular functions, and therefore is also a
Fourier transform. Let q be the summable function whose transform is this trapezoidal function. Define $q_n(x) = n^{-1}q(x/n)$ for
positive integers n. Then for each n, $||q_n||_1 = ||q||_1$. The transform
of q_n is $\hat{q}(nx)$, another trapezoidal function, equal to 1 on the
interval $(-1/n, 1/n)$.

*For any fixed positive t, $||T_t q_n - q_n||_1$ tends to 0 as n tends
to ∞.* We have

$$(2.1) \qquad ||T_t q_n - q_n||_1 = n^{-1}\int_{-\infty}^{\infty} |q((x-t)/n) - q(x/n)\, dx =$$

$$\int_{-\infty}^{\infty} |q(x - t/n) - q(x)|\, dx.$$

This quantity tends to 0 because translation is continuous in
$\mathbf{L}^1(\mathbf{R})$.

(e) *Suppose that f is in $\mathbf{L}^1(\mathbf{R})$, $\hat{f}(0) = 0$, and $\epsilon > 0$. There is
a summable function g such that $\hat{g}$ vanishes on a neighborhood of
0, and $||f - g||_1 < \epsilon$.*

It will suffice to show that any linear functional vanishing
on all g such that $\hat{g}$ is 0 on a neighborhood of 0 must vanish also
on f. Let φ be a bounded function with $\varphi * g(0) = 0$ for all such g;
we shall show that φ is constant (and thus orthogonal to f).

First we establish that $\varphi * q_n = \varphi$ for each of the trapezoidal
functions q_n of (d). Let k be any summable function. Then

(2.2) $$(\varphi * q_n - \varphi) * k = \varphi * (q_n * k - k).$$

The second factor on the right side has transform vanishing on a neighborhood of 0; by hypothesis φ annihilates such functions. Hence the left side is 0 for all k; it follows that $\varphi * q_n = \varphi$, as we wished to prove.

Now fix positive numbers t and ϵ. We have

$$(2.3) \quad ||T_t\varphi - \varphi||_\infty = ||T_t(\varphi * q_n) - \varphi * q_n||_\infty = ||\varphi * (T_t q_n) - \varphi * q_n||_\infty$$

$$\leq ||T_t q_n - q_n||_1 \, ||\varphi||_\infty \leq \epsilon ||\varphi||_\infty$$

if n is large enough. Hence $T_t\varphi = \varphi$. This equality holds a.e. for each t; it follows (after some reflection) that φ is a.e. equal to a constant.

If we set $h = f - g$, then $\hat{h} = \hat{f}$ on a neighborhood of 0, and $||h||_1 < \epsilon$. This is the function we shall need.

(f) This assertion will not be used to prove the Tauberian theorem, but it is an interesting corollary of (e). *If f is summable and ϵ positive, there is a function g in $\mathbf{L}^1(\mathbf{R})$ such that $\hat{g} = \hat{f}(0)$ on a neighborhood of 0, and $||f - g||_1 < \epsilon$.*

Let q be the trapezoidal function again. Then $f - \hat{f}(0)q$ is summable and has transform vanishing at 0. By (e), we can find h with transform vanishing *near* 0, such that $||f - \hat{f}(0)q - h||_1 < \epsilon$. Then $g = h + \hat{f}(0)q$ has the required properties.

(g) *Suppose that f is summable and $\hat{f}(0) \neq 0$. Then there is a summable g such that $\hat{f}\hat{g} = 1$ on a neighborhood of 0.*

Say $\hat{f}(0) = 1$. By (e) (applied to the function $q - f$), there is an h so that $\hat{h} = 1 - \hat{f}$ on a neighborhood of 0, and $||h||_1 < 1$. Then on a neighborhood of 0

$$(2.4) \qquad \hat{f}^{-1} = 1 + \hat{h} + \hat{h}^2 + \dots.$$

The series

$$(2.5) \qquad g = q + h + h*h + \dots$$

converges in norm, and the transform of g is the right side of (2.4) near 0. This proves the assertion.

(h) *If I and J are intersecting open intervals on the line and $\hat{f}\hat{g} = 1$ on I, $\hat{f}\hat{h} = 1$ on J, then there is a summable function k such that $\hat{f}\hat{k} = 1$ on $I \cup J$.*

Let $I = (a, c)$, $J = (b, d)$ with $a < b < c < d$. Define p as the function whose transform is the trapezoidal function equal to 1 on (a, b), decreasing to 0 on (b, c), and also to the left of a. Let q be the function similarly associated with J: $\hat{q} = 1$ on the part of J not in I, increases from 0 to 1 on (b, c), and decreases to 0 to the right of d. The crucial point is that $\hat{p} + \hat{q} = 1$ on $I \cup J$. Therefore $f*(p*g + q*h)$ has transform equal to 1 on $I \cup J$, so that $k = p*g + q*h$ is the function we need.

Now we can complete the proof of (c), and of the Tauberian theorem. Translating the result of (g), for each point x of the given interval $[a, b]$ we can find g in $\mathbf{L}^1(\mathbf{R})$ such that $\hat{f}\hat{g} = 1$ on a neighborhood of x. (This is also true of the endpoints a and b.) Let $I_1, \dots, I_n$ be a finite set of such open intervals whose union covers $[a, b]$. Name the intervals so that a belongs to I_1, the right endpoint of I_1 is in I_2, and so on. By (h) there is a function whose transform is inverse to $\hat{f}$ on all of $[a, b]$, and this finishes the proof.

A *Tauberian theorem* asserts that if a series or integral is summable in some sense, then under hypotheses it actually converges (or is summable in another sense). This is the opposite of an *Abelian theorem*, which says that a convergent series or integral is summable in some sense. Wiener viewed his theorem as an abstract Tauberian theorem, and he showed that it implies a number of previously known concrete Tauberian theorems, although the deduction was not always simple.

Here is an idea of the connection. Let φ be a bounded function that may not have a limit at ∞, but such that $\varphi * f$ has a limit for a certain summable function f. In other words, certain averages of φ have a limit. The set of all summable g such that $\varphi * g$ has a limit is a closed translation-invariant subspace of $\mathbf{L}^1(\mathbf{R})$; therefore every g in $\mathbf{M}_f$ has this property. If the Fourier transform of f is never 0, then $\mathbf{M}_f$ is all of $\mathbf{L}^1(\mathbf{R})$, and this is the conclusion that Wiener wanted.

Problems

1. Prove (a) above.

2. Show that if φ is a function such that $T_t\varphi = \varphi$ a.e. for each real t, then φ is a.e. equal to a constant.

3. Extract from the proof of the Tauberian theorem a proof that the inverse of a nonvanishing absolutely convergent Fourier series has absolutely convergent Fourier series.

4. Let $\mathbf{M}$ be a proper closed ideal of $\mathbf{L}^1(\mathbf{R})$. Show that the transforms of the functions of $\mathbf{M}$ have a common zero.

5. Let φ be a bounded uniformly continuous function in $\mathbf{H}^\infty(\mathbf{R})$. Suppose that its analytic extension to the upper

halfplane, $\Phi(x + iy)$, has a limit as x tends to ∞, for some positive y. Show that $\varphi(x)$ has a limit as x tends to ∞.

3. Spectral sets of bounded functions

Bounded functions on $\mathbf{R}$ do not generally have Fourier transforms in any sense that we have yet described, but there are related operations that transform bounded functions into some object more general than a function. The simplest idea is to observe that multiplication by a bounded function is a linear operator in $\mathbf{L}^2(\mathbf{R})$. Define an operator T in $\mathbf{L}^2(\mathbf{R})$ by setting

$$(3.1) \qquad\qquad (Tf)\hat{} = \varphi \hat{f}.$$

Then T is linear and commutes with translation. If φ is a Fourier-Stieltjes transform, T is convolution by a measure. So in general T can be thought of as a generalized convolution operator whose inverse transform is φ.

Another idea is to regard φ as a linear functional by integration on the space of infinitely differentiable functions that tend to 0 faster than any (negative) power of the independent variable. This space of functions is carried onto itself by the Fourier transform; hence the transform of a bounded function is another functional on this space. This is the point of view of the Theory of Distributions of L. Schwartz.

Although the transform of a bounded function φ cannot always be a function, it is possible to tell what the *closed support* of $\hat{\varphi}$ ought to be. This set is called the *spectral set* or the *spectrum* of φ. It was defined and studied on the real line by A. Beurling, and on locally compact abelian groups by H. Cartan and R. Godement.

The spectral set $\sigma(\varphi)$ of the bounded function φ is the set
of real λ such that whenever f is in $\mathbf{L}^1(\mathbf{R})$ and $f*\varphi$ is identically 0,
necessarily $\hat{f}(\lambda) = 0$.

Equivalently: $\sigma(\varphi)$ consists of all λ such that $\exp i\lambda x$
belongs to the weak* closure of the span of all translates of φ in
$\mathbf{L}^\infty(\mathbf{R})$.

The two criteria are easily compared. The condition $f*\varphi$
$= 0$ means that f is orthogonal to all translates of φ (when $\mathbf{L}^1(\mathbf{R})$
and $\mathbf{L}^\infty(\mathbf{R})$ are paired in the proper way). And $\hat{f}(\lambda) = 0$ just if f is
orthogonal to $\exp i\lambda x$ in this pairing. By Banach's criterion, the
exponential belongs to the *-closure of the linear span of trans-
lates of φ if and only if every f in $\mathbf{L}^1(\mathbf{R})$ that vanishes on all
translates of φ also vanishes on the exponential. This is the equivalence of the two definitions.

The first definition is the basic one. The properties of
spectral sets are derived in a series of steps.

(a) *The spectral set of the exponential* $\exp i\lambda x$ *consists of
the single point* λ. For trivially λ belongs to the spectral set. In
the other direction, if τ is different from λ, there is a summable
function f such that $\hat{f}(\lambda) = 0$ but $\hat{f}(\tau) \neq 0$ ($\hat{f}$ can be a triangular
function centered at τ with small base), so τ is not in the spectral
set.

Further, the spectral set of a trigonometric polynomial
consists exactly of the frequencies that appear in its representation. This justifies the idea of the spectral set as the set of frequencies that should appear in a harmonic analysis of φ.

(b) If φ is both summable and bounded, and f summable,
then $f*\varphi = 0$ if and only if $\hat{f}\hat{\varphi} = 0$. That is, $\hat{f} = 0$ on the support of

$\hat{\varphi}$. Hence $\sigma(\varphi)$ *is precisely the closed support of* $\hat{\varphi}$. Thus it is reasonable to think of $\sigma(\varphi)$ as a substitute for the closed support of the Fourier transform of φ.

(c) $\sigma(\varphi)$ *is closed.* By definition, it consists of the common zeros of the summable functions $\hat{f}$ such that f is orthogonal to φ and its translates; since the $\hat{f}$ are continuous, the set is closed.

(d) *For bounded functions* $\varphi,$ ψ *we have* $\sigma(\varphi + \psi) \subset \sigma(\varphi) \cup \sigma(\psi)$. For suppose λ is not in the union. Find f and g in $\mathbf{L}^1(\mathbf{R})$ such that $f*\varphi = 0$, $g*\psi = 0$, but $\hat{f}(\lambda)$ and $\hat{g}(\lambda) \neq 0$. Then $f*g*(\varphi + \psi) = 0$ but $(f*g)\hat{\ }(\lambda) \neq 0$. Thus λ is not in $\sigma(\varphi + \psi)$.

Also, $\sigma(k\varphi) = \sigma(\varphi)$ for any nonzero number k.

(e) *If f is summable and φ bounded, $\sigma(f*\varphi)$ is contained in the intersection of $\sigma(\varphi)$ with the closed support of* $\hat{f}$. This follows from the definition.

(f) *If E is a closed subset of* $\mathbf{R}$, *the set of bounded functions* φ *such that* $\sigma(\varphi) \subset E$ *constitutes a subspace of* $\mathbf{L}^\infty(\mathbf{R})$ *that is invariant under translation.* This also follows from the definition. We shall see presently that this subspace is weak* closed, but that is not obvious.

These elementary properties are to be distinguished from others that depend on the Tauberian theorem.

(g) *If $\sigma(\varphi)$ is empty, then $\varphi = 0$.* This is the statement about $\mathbf{L}^\infty(\mathbf{R})$ that is dual to the Tauberian theorem. Suppose that φ is bounded and has empty spectral set. Denote by $\mathbf{M}$ the set of all f in $\mathbf{L}^1(\mathbf{R})$ such that $f*\varphi = 0$. This is obviously a closed ideal. By the definition of spectral set, the Fourier transforms of functions in $\mathbf{M}$ have no common zero. Problem 4 of the last section shows that $\mathbf{M}$ is all of $\mathbf{L}^1(\mathbf{R})$, so that $\varphi = 0$.

Conversely, suppose that we know the assertion (g) and f is a summable function whose transform never vanishes. If φ is bounded and vanishes as a linear functional on all translates of f, that is if $f*\varphi = 0$, then $\hat{f}$ vanishes on $\sigma(\varphi)$. Hence $\sigma(\varphi)$ is empty, so $\varphi = 0$. This shows that translates of f span a dense subspace of $\mathbf{L}^1(\mathbf{R})$, which is the conclusion of the Tauberian theorem.

(h) *If f is summable and $\hat{f} = 0$ on a neighborhood of $\sigma(\varphi)$, then $f*\varphi = 0$.* For the closed support of $\hat{f}$ is disjoint from $\sigma(\varphi)$. By (e) above, $f*\varphi$ has empty spectral set, and so is null.

This assertion is a partial converse to the definition of spectral set: necessarily $\hat{f} = 0$ on $\sigma(\varphi)$ if $f*\varphi = 0$, and now it is shown that the convolution vanishes if $\hat{f} = 0$ *on a neighborhood of* $\sigma(\varphi)$. The question whether $f*\varphi = 0$ if merely $\hat{f} = 0$ on $\sigma(\varphi)$ itself was the celebrated *problem of spectral synthesis* posed by Beurling. A negative answer was given for the analogous problem in space of three dimensions (or greater) by L. Schwartz in 1947, and for Beurling's problem by P. Malliavin in 1959.

The facts can be restated in this way: if V is any neighborhood of $\sigma(\varphi)$, then φ is contained in the $*$-closed subspace of $\mathbf{L}^\infty(\mathbf{R})$ generated by exponentials $\exp i\lambda x$ with λ in V; but frequencies λ from $\sigma(\varphi)$ may not suffice.

(i) *The set of all bounded functions with spectral set contained in a given closed set E is $*$-closed.* Let λ be any point not in E. Find f in $\mathbf{L}^1(\mathbf{R})$ such that $\hat{f}$ is null on a neighborhood of E, but not at λ. Then $f*\varphi = 0$ for all φ with spectral set contained in E, and also for any ψ that is a $*$-limit of such functions. Therefore λ is not in the spectral set of ψ, and this proves the assertion.

(j) *For any bounded functions φ, ψ, the spectral set of $\varphi\psi$ is contained in the closure of $\sigma(\varphi) + \sigma(\psi)$* (the set of all $r + s$

where r is in $\sigma(\varphi)$ and s in $\sigma(\psi)$).

If φ is a trigonometric polynomial, the fact is easy to prove. Let V_n be the neighborhood of $\sigma(\varphi)$ consisting of all points whose distance from $\sigma(\varphi)$ is less than $1/n$. Find a net (φ_α) of trigonometric polynomials converging in the star topology to φ, such that each φ_α has frequencies in V_n (with n fixed). Then $\varphi_\alpha\psi$ has spectral set in $V_n + \sigma(\psi)$; and the product converges weak* to $\varphi\psi$. By (i), $\varphi\psi$ has spectral set contained in the closure of $V_n + \sigma(\psi)$. The intersection of these sets, as n tends to ∞, is the closure of $\sigma(\varphi) + \sigma(\psi)$, and this proves the assertion.

(k) **Primary Ideal Theorem.** *If φ has exactly one point λ in its spectral set, then φ is a constant times $\exp i\lambda x$.*

After multiplying φ by an exponential we can suppose that λ is 0. The conclusion is then that φ is constant. Given the Tauberian theorem, the statement is a corollary of assertion (e) of the last section. Let $\mathbf{M}$ be the closed ideal of all f in $\mathbf{L}^1(\mathbf{R})$ such that $f*\varphi = 0$. By (h), $\mathbf{M}$ contains all f such that $\hat{f}$ vanishes on a neighborhood of 0. But (e) of the last section implies then that $\mathbf{M}$ contains all f such that $\hat{f}$ vanishes *at* 0. Thus φ, as a linear functional on $\mathbf{L}^1(\mathbf{R})$, has the same null space as integration. It follows that φ is constant.

The dual statement reads: *a closed ideal in $\mathbf{L}^1(\mathbf{R})$ whose Fourier transforms have only one common zero necessarily contains all functions whose transforms vanish at that point.* In other words, every closed primary ideal is maximal. The theorem was probably proved first by V. Ditkin, and the generalization to locally compact abelian groups is due to I. Kaplansky.

Finally, we give a function-theoretic characterization of spectral set, due to Beurling but related to older work of T.

Carleman. We prove the result for bounded sequences, rather than for functions on the line.

The definition of the spectral set of a bounded sequence is analogous to that for bounded functions on $\mathbf{R}$, but the spectral set is a subset of $\mathbf{T}$ instead of $\mathbf{R}$. If α is a bounded sequence, its spectral set $\sigma(\alpha)$ is the set of all $\exp i\lambda$ such that $\beta * \alpha = 0$ (with β in $\mathbf{l}^1$) implies that the Fourier transform of β vanishes at $\exp i\lambda$:

$$(3.2) \qquad\qquad \sum \beta_n e^{-ni\lambda} = 0.$$

The properties (a – k) above have analogues for spectral sets of sequences.

Let α be any bounded sequence. The spectral set of α ought to be the complement of the open set in $\mathbf{T}$ where the Fourier transform

$$(3.3) \qquad\qquad \sum_{-\infty}^{\infty} \alpha_n e^{-nix} \quad \text{or} \quad \sum_{-\infty}^{\infty} \alpha_{-n} e^{nix}$$

vanishes. Define two analytic functions, one inside the disk and the other outside:

$$(3.4) \quad F(z) = \sum_{0}^{\infty} \alpha_{-n} z^n \quad (|z| < 1), \quad G(z) = -\sum_{-\infty}^{-1} \alpha_{-n} z^n \quad (|z| > 1).$$

If F and G have boundary values, and are equal, on an arc of $\mathbf{T}$, this is a way of saying that the series (3.3) vanish on the arc. In that case, with some growth condition, F and G are analytic and continue each other across the arc. The following theorem is a complete and surprising realization of this idea.

Theorem 33. *If J is an open arc of the circle disjoint from*

$\sigma(\alpha)$, *then F and G continue each other across J. If J contains a*
point of $\sigma(\alpha)$, then F and G do not continue each other across J.

Define a sequence γ in l^2 by setting $\gamma_n = (\alpha_n - \alpha_0)/n$. The
value of γ_0 is for the moment unspecified. Set $g = \hat{\gamma}$. Then

$$(3.5) \qquad \alpha_n = \alpha_0 + n \int g \chi^n \, d\sigma.$$

This formula holds for $n = 0$ whatever value is assigned to γ_0.

If f is a smooth function with support in a compact subset
of J, with Fourier sequence (β_{-n}) (so that the Fourier transform
of β is f), then $\alpha*\beta = 0$ by (h) above (restated for the integer
group). This means that

$$(3.6) \qquad 0 = \sum \alpha_n \beta_{-n} = \alpha_0 \sum \beta_{-n} - i \int_J g f' \, d\sigma.$$

We can arrange that the point 1 of $\mathbf{T}$ is outside J; then $\sum \beta_{-n} =$
$f(1) = 0$. Hence the integral vanishes. This fact, valid for all such
functions f, implies (Problem 4 below) that g must be constant on
J. The mean value γ_0 is at our disposal; we choose it so that $g = 0$
on J.

If we change n to $-n$ in (3.5), multiply by z^n, $|z| < 1$, and
sum we find

$$(3.7) \quad F(z) = \frac{\alpha_0}{1-z} - \int \sum_1^\infty n z^n \chi^{-n} g \, d\sigma = \frac{\alpha_0}{1-z} - \int \frac{z \bar\chi g}{(1 - z \bar\chi)^2} \, d\sigma.$$

Since g vanishes on J, this integral represents a function that is
analytic across J and everywhere outside the closed disk.

Divide numerator and denominator of the integrand on the
right side by $(z\bar\chi)^2$ and expand in a series of powers of $1/z\bar\chi$; we

find $G(z)$. This proves that F and G continue each other across every arc of **T** that is disjoint from $\sigma(\alpha)$.

Let us show that if J contains a point of $\sigma(\alpha)$, F and G do not continue each other across J. Assume that they do. Let λ be a point of $\sigma(\alpha)$ inside J. Find a sequence β in $\mathbf{l}^1$ such that $\hat{\beta} = f$ is not 0 at λ, with the support of $\hat{\beta}$ contained in an arc strictly smaller than J. Then $\alpha*\beta$ is not the null sequence (by the definition of spectral set). After translating β if necessary, $\alpha*\beta(0) \neq 0$. Write β as the inverse Fourier transform of f:

$$(3.8) \quad \sum_{-\infty}^{\infty} r^{|n|}\alpha_{-n}\beta_n = \int\left(\sum_{0}^{\infty}\alpha_{-n}r^n + \sum_{-\infty}^{-1}\alpha_{-n}r^{|n|}\right)f\chi^n \, d\sigma$$

$$= \int [F(re^{ix}) - G(r^{-1}e^{ix})]f(e^{ix}) \, d\sigma(x).$$

The bracket tends uniformly to 0 as r increases to 1, because F and G continue each other across all of J, and the support of f is a smaller interval. But the first sum tends to $\alpha*\beta(0) \neq 0$. The contradiction completes the proof.

An analogous theorem is true on the real line.

Problems

1. Show that if f is in $\mathbf{L}^1(\mathbf{R})$, φ in $\mathbf{L}^\infty(\mathbf{R})$, and $\hat{f} = 1$ on a neighborhood of $\sigma(\varphi)$, then $f*\varphi = \varphi$.

2. If φ is bounded and $\sigma(\varphi)$ compact, then φ is almost everywhere equal to the restriction of an entire function to the real axis.

3. Suppose that φ is bounded and uniformly continuous, and $\sigma(\varphi)$ consists of a discrete set of points without finite limit point. Show that φ is almost periodic; that is, φ is the uniform

limit of a sequence of trigonometric polynomials. [The result is due to Beurling. If (e_n) is any approximate identity, then $e_n * \varphi$ tends to φ uniformly.]

4. Prove the statement used above: if in (3.6) the integral vanishes for every continuously differentiable function f with support in J, then g is constant on J.

4. A theorem of Szegö; the theorem of Grużewska and Rajchman; idempotent measures

Szegö's Theorem. *Let (α_n) be a sequence whose terms are drawn from a finite set W of complex numbers. Unless the sequence is periodic, its spectral set fills* **T**.

This is not quite Szegö's statement (see Problem 2 below), but it embodies much of the content of his theorem. Problem 1 below asks for a description of the spectral set of a periodic sequence.

We assume that the spectral set E of (α_n) is not all of **T** and that the sequence is not periodic. These assumptions will lead to a contradiction.

Denote by **L** the $*$-closed subspace of $\mathbf{l}^\infty$ spanned by the translates of α. The sequences in **L** have spectral sets contained in E. We shall show that **L** contains a one-sided sequence γ: $\gamma_0 \neq 0$ but $\gamma_n = 0$ for all $n > 0$ or else for all $n < 0$. Then we shall show that the spectral set of such a sequence must fill the circle.

A *pattern of length* k will be a sequence $(q_1, \dots, q_k)$ of elements of W. There are only finitely many patterns of length k; therefore one, at least, must recur infinitely often in the sequence (α_n). By translating this sequence and subtracting, we obtain a new sequence with k consecutive 0's. Since (α_n) is not periodic, this difference is not the null sequence, so the string of 0's ends on one side at least with a nonzero element of the difference set of W. After translation, this element has index 0. We construct such a difference of translates for each positive integer k, and find a weak* accumulation point for this sequence. This is a one-sided sequence (γ_n), belonging to $\mathbf{L}$ and therefore having spectral set contained in E.

The fact that a bounded one-sided sequence has full spectral set belongs to the family of theorems asserting that analytic functions in the disk cannot vanish on a large subset of the boundary without vanishing identically. Here the analytic function merely has bounded Taylor coefficients, which is less than belonging to $\mathbf{H}^1(\mathbf{T})$; but the boundary object is null on a whole arc of the circle (because E is closed), rather than merely on a set of positive measure.

Say $\gamma_n = 0$ for all $n < 0$. Find a sequence (ρ_n) in $\mathbf{l}^1$ such that $\rho_0 \neq 0$, with Fourier transform supported on a small neighborhood V of the identity in $\mathbf{T}$. From property (j) of the last section, $(\gamma_n \rho_n)$ is a sequence whose spectral set is contained in the closure of $E + V$. If V is small enough, this is still not the whole of $\mathbf{T}$. Moreover the sequence is in $\mathbf{l}^1$, so its spectral set is the closure of the support of its Fourier transform (property (b)). Therefore the conjugate-analytic function

$$(4.1) \qquad \sum_{0}^{\infty} \gamma_n \rho_n e^{-inx}$$

vanishes on an arc of $\mathbf{T}$, but not identically. This is impossible, and the contradiction proves the theorem.

The theorem can be improved. Let (α_n) be a bounded

sequence, and I an open arc of $\mathbf{T}$. Say that $\sigma(\alpha)$ is *measurable on* I if there is a measure μ on $\mathbf{T}$ such that the spectral set of $(\alpha_n - a_n(\mu))$ does not intersect I. The stronger result (due to the author) is

Theorem 34. *If (α_n) has terms drawn from a finite set, and if its spectral set is measurable on some open arc I of $\mathbf{T}$, then the sequence is ultimately periodic to the right, and also to the left.*

Suppose that (α_n) is not ultimately periodic to the right. (The proof for the other case is similar.) Then there are strictly increasing sequences (m_k), (n_k) of positive integers such that

$$(4.2) \qquad \alpha_{m_k+j} = \alpha_{n_k+j}$$

for $j = 0, \ldots, k-1$, but the equality is false for $j = k$. Define the left-shift: $(S\alpha)_n = \alpha_{n+1}$, and the sequences

$$(4.3) \qquad \gamma^k = S^{m_k+k}\alpha - S^{n_k+k}\alpha.$$

Then γ^k vanishes at indices $-1, \ldots, -k$, but γ^k_0 is a nonzero element of the difference set of W. A subsequence of (γ^k) converges pointwise, and thus in the weak∗ topology of l^∞, to a sequence γ that vanishes to the left, but not at 0. So far, we are repeating the proof of Szegő's theorem.

Let μ be a measure on $\mathbf{T}$ such that $(\alpha_n - a_n(\mu))$ has spectral set disjoint from I. Exceptionally, let $a_n(\mu)$ mean $\hat{\mu}(-n)$ (this is still a Fourier-Stieltjes sequence). Apply the translation operations of (4.3) to $(a_n(\mu))$. The corresponding sequence a^k is the Fourier-Stieltjes sequence of the measure $(\chi^{m_k+k} - \chi^{n_k+k})\mu$, which we call μ_k. We passed to a subsequence of k to have convergence of (γ^k) to γ; on a further subsequence, μ_k converges weak$*$ to a measure ν on **T**.

Lemma. ν *is singular with respect to* σ.

Assume the lemma for the moment in order to finish the proof. Now $\gamma - a(\nu)$ is a sequence whose spectral set is still disjoint from I, and $\gamma_n = 0$ for $n < 0$, $\gamma_0 \neq 0$. Let β be the inverse Fourier sequence of a triangular function f with support in an interval smaller than I. By (h) of the last section, $\beta*(\gamma - a(\nu)) = 0$. Since $\beta_n = O(1/n^2)$ for $n \to \pm\infty$, and γ is one-sided, $\beta*\gamma(n) = O(1/n)$ for $n \to -\infty$. Hence the same is true for $\beta*a(\nu) = \beta*\gamma$. Thus the measure $f\,d\nu$, which has inverse Fourier-Stieltjes sequence $\beta*a(\nu)$, is the sum of a square-summable function and a measure whose coefficients vanish on one side. By the theorem of F. and M. Riesz, $f\,d\nu$ is absolutely continuous; by the lemma it is singular, so $f\,d\nu = 0$. It follows that $\beta*a(\nu) = 0$, and hence that $\beta*\gamma = 0$. This means that the spectral set of γ is disjoint from the support of f. But we know that the spectral set of a nonnull one-sided sequence fills **T**. This contradiction means that α must have been periodic to the right.

The lemma remains. Write $\mu = \mu_a + \mu_s$ where μ_a is absolutely continuous and μ_s is singular. Since the sequences (m_k) and (n_k) tend to ∞, $(\chi^{m_k+k} - \chi^{n_k+k})\mu_a$ tends to the null measure in the weak$*$ topology of measures, by Mercer's theorem. Therefore we only have to show that the limit ν of $(\chi^{m_k+k} - \chi^{n_k+k})\mu_s$ (on a subsequence of k) is singular. To simplify notation, write μ for μ_s and suppose that this sequence converges weak$*$ to ν.

A positive measure ζ on **T** is singular if and only if for each positive ϵ there is an open set G such that $\sigma(G) = 1$, $\zeta(G) < \epsilon$. Find such a set for $|\mu|$. For each function f that is continuous on $\mathbf{T}$, supported on G, and has $||f||_\infty \le 1$, we have

(4.4) $\left| \int f d\nu \right| = \lim \left| \int f(\chi^{m_k+k} - \chi^{n_k+k}) \, d\mu \right| \le 2\epsilon.$

One can show, using the theorem of F. Riesz, that the supremum over f of the left side is $|\nu|(G)$. The criterion above shows that ν is singular, and the proof is finished.

A complex measure μ on $\mathbf{T}$ is *idempotent* if $\mu*\mu = \mu$. This is the case if and only if the Fourier-Stieltjes coefficients of μ are all equal to 0 or 1. Which sequences of 0's and 1's are Fourier-Stieltjes sequences? The answer (also due to the author) follows from the theorem just proved. The spectral set of $(a_n(\mu))$ is measurable on all of $\mathbf{T}$, and therefore the sequence must be ultimately periodic to the right and to the left. This condition is not quite sufficient.

Theorem 35. *A sequence of 0's and 1's is a Fourier-Stieltjes sequence if and only if it coincides with a periodic sequence except in finitely many terms.*

A periodic sequence is always a Fourier-Stieltjes sequence, and changing finitely many terms only means adding a trigonometric polynomial to the measure. In the other direction there is still a point to be established: a Fourier-Stieltjes sequence that is periodic to the left and to the right must coincide with a single periodic sequence except in finitely many terms. This will follow from the

Theorem of Gruzewska and Rajchman. *If μ is a measure on $\mathbf{T}$ and $a_n(\mu)$ has a limit as $n \to -\infty$, then the sequence has the same limit as $n \to \infty$.*

By subtracting a point mass at 0 from μ we can suppose

that the limit at $-\infty$ is 0. If P is a trigonometric polynomial, it is obvious that $a_n(P\,d\mu)\to 0$ as $n\to -\infty$. Trigonometric polynomials are dense in the space $\mathbf{L}^1$ formed with the positive measure $|\mu|$; it follows that any measure absolutely continuous with respect to μ has the same property. Hence the complex conjugate measure $\bar{\mu}$ has coefficients tending to 0 at $-\infty$. This means that the coefficients of μ itself tend to 0 as n tends to ∞, as we wished to prove.

Now finish the proof of Theorem 35. If the theorem were not true, we could obtain by translation and subtraction a Fourier-Stieltjes sequence vanishing on one side, but not tending to 0 on the other; but this is impossible.

The theorem on idempotent measures was generalized to torus groups by W. Rudin, and to arbitrary compact abelian groups by P. J. Cohen.

Problems

1. Describe the spectral set of a periodic sequence; show that a periodic sequence is a Fourier-Stieltjes sequence.

2. Prove Szegő's theorem, as he stated it: *if a power series has only finitely many distinct coefficients, and if its sum f has an analytic continuation across some arc of the unit circle, then f is a rational function with poles at roots of unity.*

3. Give an example of a sequence that is ultimately periodic to the left, and to the right, but that is not a Fourier-Stieltjes sequence.

4. Describe (as measures) the idempotent measures on $\mathbf{T}$.

5. Prove the corollary of Riesz' theorem used above.

Chapter 7
Distribution

1. Equidistribution of sequences

For any real number u let $[u]$ denote the integer part of u, and (u) the fractional part. Then $u = [u] + (u)$, where the first summand is an integer, and the second satisfies $0 \le (u) < 1$. If u is irrational, the fractional parts (ku), $k = 1, 2, \ldots$ move about the interval $(0, 1)$ in what seems at first to be a random way, but in fact a good deal can be said about their distribution. In this chapter we shall prove theorems about the distribution of this and other sequences in the interval. The definition of equidistribution and the results of this chapter are due to H. Weyl. The methods and results presented here were chosen because they involve trigonometric series.

The notation (u_k) can mean either the sequence whose elements are u_k, or the fractional part of a number u_k. We shall continue to mean the former unless otherwise stated.

Given a sequence (u_k) of real numbers, we can study the distribution of their fractional parts by forming the sequence $(\exp 2\pi i u_k)$ of numbers on $\mathbf{T}$. This makes available the analytic properties of the exponential function.

Let (u_k) $(k \ge 1)$ be a sequence of real numbers contained in an interval I of length $|I|$. For any subinterval J of I, of length $|J|$, let $J(n)$ denote the number of points among $u_1, \ldots, u_n$ that lie in J. The sequence is said to be *equidistributed* or *uniformly distributed* on I if for each J contained in I

$$(1.1) \qquad \lim_n \frac{J(n)}{n} = \frac{|J|}{|I|}.$$

(Intervals may be open or closed or half-open.) In words: the asymptotic proportion of terms of the sequence falling in J is the normalized length of J, for each subinterval J of I.

The definition is geometrically simple but not easy to use analytically. It is translated to analysis by this fundamental result.

Theorem 36. *The sequence* (u_k) *contained in* $[0, 2\pi)$ *is uniformly distributed on that interval if and only if*

$$(1.2) \qquad \lim_n n^{-1} \sum_1^n f(u_k) = \int f \, d\sigma$$

for every function f that is continuous and periodic with period 2π. In order for this to be the case, it is sufficient that (1.2) hold for the functions $\exp jit$, *that is,*

$$(1.3) \qquad \lim_n n^{-1} \sum_{k=1}^n e^{jiu_k} = 0 \qquad (\text{all integers } j \neq 0).$$

The proof will depend on this lemma.

Lemma. *If (1.2) holds for each of a sequence of functions* (f_j), *and if f_j tends to f uniformly, then (1.2) is true for f.*

The simple proof is omitted.

The definition of equidistribution is simply (1.2) for each f that is the characteristic function of a subinterval J of $[0, 2\pi)$. Suppose that (1.2) holds for such functions. Any continuous, periodic function can be approximated uniformly by step functions, that is, by linear combinations of characteristic functions of intervals. By the lemma, (1.2) holds for the limit function.

Conversely, suppose that (1.2) holds for all continuous periodic f. Let g be the characteristic function of a subinterval J

of $[0, 2\pi)$. Construct continuous periodic functions h and k such that $h \leq g \leq k$ and the integral of $k - h$ is small. The sums (1.2) for g lie between the corresponding sums for h and k. Thus we have

$$(1.4) \qquad \int h \, d\sigma \leq \liminf n^{-1} \sum_1^n g(u_k) \leq \limsup n^{-1} \sum_1^n g(u_k) \leq \int k \, d\sigma.$$

Since the left and right members are arbitrarily close together, the limit (1.2) exists for g and the equality is proved. We have now shown that (1.2) for all continuous periodic functions f is necessary and sufficient for the uniform distributiion of (u_k).

If (1.2) holds for all continuous periodic functions f, then of course (1.3) is true. In the other direction, (1.3) implies that (1.2) holds for trigonometric polynomials, and so (1.2) follows for all continuous periodic functions by the lemma. This completes the proof.

Corollary. *If u is irrational, then the fractional parts (ku) are uniformly distributed on $(0, 1)$.*

We have to verify that

$$(1.5) \qquad \lim n^{-1} \sum_{k=1}^n e^{2\pi i j k u} = 0 \qquad (\text{all } j \neq 0).$$

The sum is a geometric series, equal to

$$(1.6) \qquad e^{2\pi i j u} \frac{1 - e^{2\pi i j n u}}{1 - e^{2\pi i j u}} \, ;$$

the denominator is not 0 for $j \neq 0$ because u is irrational. Divided by n, this quantity tends to 0.

It follows that the fractional parts (ku) are *dense* in the interval.

Theorem 36 suggests a definition of distributions on an interval other than the uniform distribution. Let μ be a probability measure on $[0, 1)$, and (u_k) a sequence contained in this interval. We say that the sequence has distribution μ if

$$(1.7) \qquad \lim_n n^{-1} \sum_1^n f(u_k) = \int f \, d\mu$$

for every continuous periodic function f. For this to be the case it is necessary and sufficient that (1.7) hold when f is an exponential $\exp 2\pi jit$, because the lemma is still true in this context. Moreover if (1.7) holds for characteristic functions of intervals, then it is true for all continuous periodic functions, but the converse is sometimes false (Problem 2 below).

Problems

1. Invent a sequence that is dense on $(0, 1)$, but not uniformly distributed.

2. Find a sequence in $(0, 2\pi)$ whose distribution measure is the unit point mass at 1, such that (1.7) fails for the characteristic function of some interval.

3. Phrase (1.7) as an assertion about weak* convergence of a sequence of measures.

4. Find a sequence on $(0, 2\pi)$ that has no distribution measure.

5. Show that the distribution measure on $\mathbf{T}$ of $\exp 2\pi i(1 + \sqrt{2})^n$ is the point mass at 1. [$1 + \sqrt{2}$ is an algebraic integer of degree 2. If u is its conjugate, then $(1 + \sqrt{2})^n + u^n$ is an ordinary integer for each positive integer n.]

6. Find the distribution of the fractional parts $(k!e)$.

2. Distribution of $(n_k u)$

From Weyl's criterion of the last section it was easy to see that the fractional parts (ku) are uniformly distributed on $(0, 1)$ for every irrational u. It is more difficult to establish the uniform distribution of other arithmetically defined sequences. In this section we prove a universal result of Weyl that puts the problem in perspective.

Theorem 37. *For any strictly increasing sequence (n_k) of positive integers, the fractional parts $(n_k u)$ are uniformly distributed on $(0, 1)$ for almost every u.*

The scale is immaterial; we prove the same fact for $(n_k u)$ modulo 2π.

Define

$$(2.1) \qquad S(p) = S(p, u) = p^{-1} \sum_{k=1}^{p} e^{j i n_k u},$$

where j is an integer different from 0 that is fixed for the moment. If for a particular u

$$(2.2) \qquad \sum_{1}^{\infty} |S(p)|^2 < \infty,$$

then $S(p)$ tends to 0, and if this is true for each $j \neq 0$ then uniform distribution is established for that u. That much is not true, but the idea will help.

For each p,

$$(2.3) \qquad \int |S(p, u)|^2 \, d\sigma(u) = 1/p.$$

Hence

(2.4) $$\int \sum |S(p^2)|^2 \, d\sigma \;=\; \sum \int |S(p^2)|^2 \, d\sigma \;<\; \infty.$$

Thus

(2.5) $$\sum_1^\infty |S(p^2, u)|^2 \;<\; \infty \quad \text{a.e.,}$$

so that $S(p^2, u)$ tends to 0 a.e.

For each positive integer m, let $p = p(m)$ satisfy $p^2 \le m < (p+1)^2$. Obviously for all u

(2.6) $$|mS(m) - p^2 S(p^2)| \;\le\; m - p^2,$$

so that

(2.7) $$|S(m) - \frac{p^2}{m} S(p^2)| \;\le\; 1 - \frac{p^2}{m},$$

a quantity that tends to 0 as m tends to ∞. The second term on the left tends to 0 for a.e. u, and thus $S(m)$ does too.

In this argument j was fixed. Outside a countable union of null sets, $S(m, j, u)$ tends to 0 for all $j \ne 0$ at once, and the theorem is proved.

Weyl himself disclaimed interest in such a theorem, which purports to tell something about almost all u but nothing about any particular u. Nevertheless it shows that we can expect uniform convergence as the normal case for sequences $(n_k u)$.

Problems

1. Let (n_k) be a strictly increasing sequence of positive integers. If $(n_k u)$ is uniformly distributed modulo 2π for all u in a

measurable set E, and if μ is a measure supported on E, then $\hat{\mu}(n_k)$ is summable to 0 by Cesàro means.

2. $S(p)$ is as in (2.1). The set of u such that $S(p, u)$ tends to a particular limit different from 0 is a Borel set. Show that it cannot support a positive measure whose Fourier-Stieltjes coefficients tend to 0.

3. $(k^r u)$

Weyl proved that if u is irrational and r any positive integer, then $(k^r u)$ is uniformly distributed modulo 1. This result is not simple, but it can be proved in a number of interesting ways. In this section two methods are presented.

The first method depends on ideas from ergodic theory. Let $\mathbf{X}$ be a compact metric space and τ a continuous mapping of $\mathbf{X}$ into itself. For x a given point of $\mathbf{X}$, we ask whether the sequence $(\tau^k x)$, $k \geq 0$, has a distribution; that is, a probability measure μ on $\mathbf{X}$ such that

$$(3.1) \qquad \lim_n n^{-1} \sum_0^{n-1} f(\tau^k x) = \int f \, d\mu$$

for every continuous function f on $\mathbf{X}$. If $\mathbf{X}$ is a compact group and μ is normalized Haar measure, we say that the sequence is uniformly distributed.

For each n let μ_n be the probability measure on $\mathbf{X}$ that has point mass $1/n$ at each point $x, \tau x, \ldots, \tau^{n-1} x$. Then (3.1) says that μ_n converges to μ in the weak$*$ topology of measures. For any x, the sequence (μ_n) is bounded in the space of measures on $\mathbf{X}$, and therefore has accumulation points for the weak$*$ topology of measures. A measure ν is an accumulation point if every weak$*$

neighborhood of ν contains infinitely many μ_n. This is the case if and only if ν is the limit of a subsequence of (μ_n).

For continuous functions f define $f^\tau(x) = f(\tau x)$, another continuous function. Dually, to each measure μ define the functional μ^τ by

$$(3.2) \qquad \mu^\tau(f) = \mu(f^\tau) = \int f^\tau \, d\mu.$$

This functional is realized by a measure, also called μ^τ. If μ is a probability measure, so is μ^τ.

Lemma. *Let x be a point of $\mathbf{X}$ and (μ_n) the measures defined above. If a subsequence (μ_{n_j}) converges weak* to μ, then $\mu^\tau = \mu$. If μ is the only probability measure on $\mathbf{X}$ with this invariance property, then μ_n converges weak* to μ, and μ is the distribution measure for $(\tau^k y)$ for every y in $\mathbf{X}$.*

For any continuous f,

$$(3.3) \qquad \mu^\tau(f) = \mu(f^\tau) = \lim_j n_j^{-1} \sum_0^{n_j-1} f^\tau(\tau^k x) = \lim_j n_j^{-1} \sum_1^{n_j} f(\tau^k x)$$
$$= \lim_j n_j^{-1} \sum_0^{n_j-1} f(\tau^k x) + \lim_j n_j^{-1}(f(\tau^{n_j} x) - f(x)).$$

The last limit is 0, and the one before is $\mu(f)$. This proves the invariance of μ.

If μ is the only invariant probability measure, then μ is the only accumulation point of (μ_n), so the sequence converges to μ. And this is true for every point y, which completes the proof.

A point x is called a *generic point* for the probability measure μ if (3.1) holds for all continuous f; or equivalently, if the measures μ_n tend to μ. The lemma shows that there is at least one probability measure invariant under τ for which a given point x is generic. If τ has only one invariant probability measure (so that every point is generic for that measure), τ is called *uniquely ergodic*. Proving that a transformation is uniquely ergodic is a way to establish equidistribution theorems. This idea is due to H. Furstenberg.

Let **X** be the circle **T**, written additively, and τ the transformation that takes x to $x + u$ (modulo 2π), where $u/2\pi$ is irrational. It is easy to prove that τ is uniquely ergodic, the invariant probability measure being Haar measure σ (Problem 1 below). Therefore the sequence $(x + ku)$ is uniformly distributed modulo 2π, for every x. If we know the result for $x = 0$ it follows easily for all x, so the added generality is not important, but this is a new proof of the Corollary of Section 1.

Now take for **X** the two-dimensional torus $\mathbf{T}^2$, represented as the set of pairs (x, y) of real numbers with addition in each coordinate modulo 2π. Let u be a real number such that $u/2\pi$ is irrational. Define a mapping of **X** into itself by setting

$$(3.4) \qquad \tau(x, y) = (x + y, y + u);$$

τ is a homeomorphism of **X**. A calculation shows that

$$(3.5) \qquad \tau^k(x, y) = (x + ky + \tfrac{1}{2}(k-1)ku, \; y + ku)$$

for each positive integer k.

The mapping τ is uniquely ergodic, the only invariant probability measure being normalized Haar measure: $d\sigma = dx\,dy/4\pi^2$.

Assume this for a moment. By the lemma, $\tau^k(x, y)$ is uniformly distributed on **X** for all x, y. Take $x = y = 0$ and $f(x, y) = \exp i(2jx + jy)$, where j is a nonzero integer. Then (3.1) gives

$$(3.6) \qquad \lim_n n^{-1} \sum_{k=0}^{n-1} \exp i(j(k-1)ku + jku) = \int f\, d\sigma = 0$$

or

$$(3.7) \qquad \lim_n n^{-1} \sum_0^{n-1} \exp ijk^2 u = 0.$$

Since j was any nonzero integer, Weyl's criterion shows that $(k^2 u)$ is uniformly distributed modulo 2π.

We want to prove that τ was uniquely ergodic. A calculation shows shows that a measure μ is invariant if and only if

$$(3.8) \qquad \hat{\mu}(m, n) = e^{-niu} \hat{\mu}(m, m+n)$$

for all integers m, n. This is obviously true if μ is Haar measure: $\hat{\mu}(m, n) = 0$ unless $m = n = 0$. We must show that (3.8) holds only for this measure.

By induction, (3.8) leads to

$$(3.9) \qquad \hat{\mu}(m, n) = \left(\exp -[\tfrac{1}{2}(k-1)km + kn]iu\right) \hat{\mu}(m, km+n)$$

for all integers m, n and positive k. The same formula holds for negative k (Problem 3 below). Now $\hat{\mu}(m, km+n)$ is a Fourier-Stieltjes sequence in the single variable k for fixed m, n (Problem 4 below). If a coefficient $\hat{\mu}(m, n)$ is not 0,

(3.10) $(\exp \frac{1}{2}(k-1)\,kmiu)\,(\exp kniu)$

would be a Fourier-Stieltjes sequence in k. We shall show that this is impossible unless $m = 0$.

Since $\exp(-kniu)$ is a Fourier-Stieltjes sequence in k, we can neglect the second factor in (3.10). Let ν be the measure such that

(3.11) $\displaystyle \int e^{-kix}\,d\nu(x) \;=\; \exp\!\left(\tfrac{1}{2}(k-1)kmiu\right)$

for all k. Then

(3.12) $(\bar{\chi}\nu)\hat{\ }(k) \;=\; \hat{\nu}(k+1) \;=\; e^{kmiu}\hat{\nu}(k),$

which is the Fourier-Stieltjes coefficient of a translate of ν through mu. Thus the coefficients of ν have constant modulus, and ν must have a point mass by the criterion of Wiener (Chapter 1, Section 4). If x carries a point mass, and if $m \neq 0$, then all the points $x + jmu$ are distinct as j ranges over the integers, and (3.12) shows that each of these points carries a mass of the same magnitude. But ν is a finite measure, so this is impossible.

Therefore $\hat{\mu}(m, n) = 0$ unless $m = 0$.

If $m = 0$ but $n \neq 0$, (3.8) shows immediately that $\hat{\mu}(0, n) = 0$. This concludes the proof.

The same idea can be used to treat $(k^r u)$ for any positive integer r.

The interest of this proof lies in the connection it creates between Diophantine problems and the part of ergodic theory

known as Topological Dynamics. Furstenberg and his collabo-
rators have developed this connection into a remarkable body of
results. Now we shall prove the same theorem in a very different
way.

Theorem of van der Corput. *A sequence* (u_k) $(k \geq 1)$ *of real
numbers is uniformly distributed modulo 1 if for every positive integer p the sequence* $(u_{k+p} - u_k)$ *is uniformly distributed modulo 1.*

The following beautiful proof is transmitted by G. Rauzy.

Let (n_k) $(k \geq 1)$ be a fixed strictly increasing sequence of
positive integers, and (α_k), (β_k) $(k \geq 1)$ two bounded complex
sequences. We define an inner product

$$(3.13) \qquad M(\alpha, \beta) = \lim_k n_k^{-1} \sum_1^{n_k} \alpha_j \bar{\beta}_j$$

if this limit exists. Since the sequences are bounded, we have for
any positive integer p

$$(3.14) \qquad M(\alpha, \beta) = \lim_k n_k^{-1} \sum_1^{n_k} \alpha_{j+p} \bar{\beta}_{j+p}.$$

If we set $\alpha_k = \beta_k = 0$ for $k \leq 0$, then (3.14) holds for all integers p.

The shift operator S is defined by $(S\alpha)_k = \alpha_{k+1}$ for positive
k, and $= 0$ for $k \leq 0$. We are interested in the sequence

$$(3.15) \qquad \rho(p) = M(S^p \alpha, \alpha),$$

assumed defined for $p = 0, 1, \ldots$ Let $\rho(-p) = \overline{\rho(p)}$, thus defining ρ
for negative p.

Lemma. *If ρ is defined for the bounded sequence α, it is
positive definite.*

The proof is formally the same as showing that $\alpha*\tilde\alpha$ is positive definite, where α belongs to 1^2.

The inner product has the expected algebraic properties:

$$(3.16) \qquad M(S^p(\alpha - \beta), (\alpha - \beta)) = M(S^p\alpha, \alpha) + M(S^p\beta, \beta)$$
$$- M(S^p\alpha, \beta) - M(S^p\beta, \alpha)$$

provided the means exist.

If $(\exp 2\pi i u_k)$ is not uniformly distributed on $\mathbf{T}$, then for some integer $r \neq 0$ and some increasing sequence (n_k)

$$(3.17) \qquad \lim_k n_k^{-1} \sum_{j=1}^{n_k} e^{2\pi i r u_j} = A \neq 0.$$

Take $\alpha_k = \exp 2\pi i r u_k$ and $\beta_k = 1$ $(k \geq 1)$. Let λ be any complex number. Using this sequence (n_k) and (3.16), we calculate $M(S^p(\alpha - \lambda\beta), (\alpha - \lambda\beta))$. The first term is

$$(3.18) \qquad M(S^p\alpha, \alpha) = \lim_k n_k^{-1} \sum_1^{n_k} e^{2\pi i r(u_{j+p} - u_j)}$$

By hypothesis, this limit is 0 for all $p > 0$, and therefore also for $p < 0$. The next term $M(S^p\lambda\beta, \lambda\beta)$ is $|\lambda|^2$ for all p. The last two terms are conjugates, with sum $2\Re(\bar\lambda A)$ for non-negative p, and therefore for all p. We know that the sum (3.16) is positive definite, and by Herglotz' theorem is the Fourier-Stieltjes sequence of a positive measure μ. Now we can identify this measure. The sequence (3.18) is the Fourier-Stieltjes sequence of σ. Denote the unit mass at the identity of $\mathbf{T}$ by δ; then the constant $|\lambda|^2$ comes from the measure $|\lambda|^2\delta$. The last terms contribute $2\Re(\bar\lambda A)\delta$. Therefore the sum

(3.19) $\mu = \sigma + (|\lambda|^2 + 2\,\Re(\bar{\lambda}A))\delta$

is a positive measure, and this holds for all complex λ. Choose λ so that $\bar{\lambda}A$ is negative. Then the coefficient of δ is negative for $|\lambda|$ small enough. But σ has no point mass, so μ is not a positive measure for such λ. This contradiction shows that (3.17) was impossible, and the theorem is proved.

Weyl's Theorem. *Let P be a real polynomial having at least one term ux^n, $n \geq 1$, with irrational coefficient u. Then the sequence $(P(k))$ is uniformly distributed modulo 1.*

We know that this is true if the polynomial has degree 1. The result follows from van der Corput's theorem by induction.

Problems

1. Show that μ^τ is a probability measure if μ is.

2. Prove that an irrational rotation on $\mathbf{T}$ is uniquely ergodic. [Compare the coefficients of μ and μ^τ.]

3. Verify that (3.9) holds for negative k.

4. Show that if $a(m, n)$ is a Fourier-Stieltjes sequence in two variables, then $a(m, km + n)$ is a Fourier-Stieltjes sequence in k for fixed m, n.

5. Carry out the inductive proof of Weyl's theorem.

APPENDIX
Integration by Parts

The formula for integration by parts

$$\int_a^b f(x)\, g'(x)\, dx = f(b)g(b) - f(a)g(a) - \int_a^b f'(x)\, g(x)\, dx$$

is proved simply by differentiating both sides with respect to b, in the ordinary case that f and g are differentiable on the interval of integration. We sometimes need a more general theorem in which $g'dx$ is replaced by $d\mu(x)$, where μ is a function of bounded variation:

Theorem. *If f is differentiable and μ is of bounded variation on $[a, b]$, then*

$$\int_a^b f(x)\, d\mu(x) = f(b)\mu(b) - f(a)\mu(a) - \int_a^b f'(x)\, \mu(x)\, dx,$$

provided that μ is continuous at the endpoints.

If μ is not continuous at the endpoints, the formula can be modified in an obvious way.

Write f as the integral of its derivative; the left side becomes

$$\int_a^b f(a) + \int_a^b f'(t)\, E(x-t)\, dt\, d\mu(x),$$

where $E(x) = 1$ for $x \geq 0$ and 0 otherwise. The term $f(a)$ integrates to $f(a)[\mu(b) - \mu(a)]$. What remains is an iterated integral to which Fubini's theorem is applicable. Interchanging the order of integration and performing the integration in x leads to

$$f(a)[\mu(b) - \mu(a)] + \int\limits_a^b [\mu(b) - \mu(t)] f'(t)\, dt.$$

When this is simplified we find the desired formula.

BIBLIOGRAPHIC NOTES

A very complete exposition of the classical theory of Fourier series with full bibliography is the treatise

A. Zygmund, *Trigonometric Series* (Second Edition), in two volumes. Cambridge University Press, 1959.

A second major work is

N. K. Bari, *A Treatise on Trigonometric Series*. Macmillan, 1964.

The encyclopedic work

E. Hewitt and K. A. Ross, *Abstract Harmonic Analysis*, in two volumes. Springer, 1963 and 1970

is modern in outlook and content.

The following texts are specifically about harmonic analysis on the classical groups:

L. Baggett and W. Fulks, *Fourier Analysis*. Anjou Press, 1979.

R. Bhatia, *Fourier Series*, Hindustan Book Agency, 1993.

R.E. Edwards, *Fourier Series, a Modern Introduction*, in two volumes, second edition. Springer, 1982.

Y. Katznelson, *An Introduction to Harmonic Analysis*. John Wiley and Sons, 1968, reprinted by Dover.

Harmonic analysis on groups other than the three classical ones is treated in

C.C. Graham and O.C. McGehee, *Essays in Commutative*

Harmonic Analysis. Springer, 1979.

L. Loomis, *Abstract Harmonic Analysis*. Van Nostrand, 1953.

W. Rudin, *Fourier Analysis on Groups*. Interscience, 1962.

The prerequisites for reading this book are contained in

W. Rudin, *Real and Complex Analysis*. McGraw Hill, 1966.

The study

J.-P. Kahane, *Séries de Fourier Absolument Convergentes*, Ergebnisse der Mathematik und ihrer Grenzgebiete, Band 50. Springer, 1970

contains both classical results and modern research.

These books are not specifically about harmonic analysis but overlap the subject:

P.L. Duren, *Theory of H^p Spaces*. Academic Press, 1970.

T.W. Gamelin, *Uniform Algebras*. Prentice-Hall, 1969.

K. Hoffman, *Banach Spaces of Analytic Functions*. Prentice-Hall, 1962.

P. Koosis, *Lectures on H_p Spaces*. Cambridge University Press, 1971.

A modern view of the Fourier transform, in which real methods are preferred to complex ones, is contained in

E.M. Stein and G. Weiss, *Introduction to Fourier Analysis on Euclidean Spaces*. Princeton University Press, 1971.

The proof of the identity at the end of Chapter 1, Section 4 is from

Robert M. Young, An elementary proof of a trigonometric identity, *Amer. Math. Monthly* 86 (1979), 296.

The theorem on convergence of Fourier series referred to in Chapter 1, Section 5 was proved in

L. Carleson, On convergence and growth of partial sums of Fourier series, *Acta Math.* 116 (1966), 135-157.

The proof of the Hausdorff-Young inequality in Chapter 1, Section 9, by A.P. Calderón and A. Zygmund is from

Contributions to Fourier Analysis, Annals of Mathematics Studies #25. Princeton University Press, 1950.

The theorem of Chapter 1, Section 10 is from

A. Beurling and H. Helson, Fourier-Stieltjes transforms with bounded powers, *Math. Scand.* 1 (1953), 120-126.

The proof of Minkowski's theorem in Chapter 3, Section 5 is from

C.L. Siegel, Über Gitterpunkte in convexen Körpern und ein damit zusammenhängendes Extremalproblem, *Acta Math.* 65 (1935), 307-323.

The two theorems of the author presented in Chapter 6, Section 4, are from

H. Helson, Note on harmonic functions, *Proc. Amer. Math. Soc.* 4 (1953), 686-691.
H. Helson, On a theorem of Szegö, *Proc. Amer. Math. Soc.* 6 (1955), 235-242.

The ideas of Furstenberg in Chapter 7 can be found in

H. Furstenberg, Strict ergodicity and transformation of the torus, *Amer. J. Math.* 83 (1961), 573-601.

The final proof in Chapter 7 is from

G. Rauzy, *Propriétés Statistiques de Suites Arithmétiques.* Presses Universitaires de France, 1976.

More information about distribution of sequences can be found in

L. Kuipers and H. Niederreiter, *Uniform Distribution of Sequences.* John Wiley and Sons, 1974.
R. Salem, *Algebraic Numbers and Fourier Analysis.* D.C. Heath and Co., 1963.

INDEX

Abelian theorem 190
absolute Fourier multiplier 176
algebra 12
approximate identity 13, 36, 55

Baggett, L. 221
Baire field 81
Banach, S. 102
Banach-Alaoglu theorem
31, 60, 71, 91
Banach-Steinhaus theorem
31, 60, 71, 91
Banach algebras 9, 12, 22
Bari, N. 221
Bessel's inequality 2, 172
Beurling, A. 9, 45, 111, 112, 118,
120, 191, 194, 195, 199, 223
bilinear form 130
Blaschke product 124, 127
Bochner's theorem 70, 90

Calderón, A.P. 43, 223
Carleman, T. 195
Carleson, L. 27, 222
Cartan, H. 195
character 79, 87
Chernoff, P. 5
Closed graph theorem 52
Cohen, P.J. 204
compact operator 90
conjugate function 134, 143, 157, 163
conjugate series 143
continuity theorem 76
convergence 5, 6, 65
convolution 11, 20, 21, 55
van der Corput 216

Diophantine approximation 99
Dirichlet kernel 25, 26

distribution function 149, 167
distribution measure 45
Ditkin, V. 195
Double series theorem 129
dual group 79
Duren, P.L. 222

Edwards, R.E. 221
endomorphism 49
equidistribution 205
event 171
extension theorem 100

factoring 113, 114
Fatou, P. 39, 106
Fejér, L. 14, 117
Fejér kernel 27, 56
Fekete, M. 62
Fourier coefficient 1
Fourier series 1
Fourier transform 7, 10, 53
Fourier-Stieltjes coefficient 19
Fourier-Stieltjes transform 7, 10, 53
fractional parts 205
Fulks, W. 221
Furstenberg, H. 213, 223

Gamelin, T.W. 222
Gelfand, I.M. 9, 12, 51
Generic point 212
Godement, R. 192
Graham, C.C. 221
Grużewska, H. 203

Haar measure 81, 86, 89
Hardy, G.H. 129, 166
Hardy spaces 105, 132, 134, 136
harmonic function 105, 132, 134, 136
Helson, H. 45, 104, 121, 203, 223

Herglotz, G. 40, 70
Hermitian form 130
Hewitt, E. 221
Hilbert, D. 129, 133
Hilbert transform 163
Hoffman, K. 222
Hunt, R. 27

ideal 13, 186
idempotent measure 203
independent random variables 171
inequality of Hausdorff
    and Young 42
inner function 113, 124
integration by parts 219
interpolation theorem 42
invariant subspace 110
inverse Fourier transform 54
inversion theorem 6, 9, 58

Jensen's inequality 121

Kahane, J.-P. 45, 222
Kaplansky, I. 195
Katznelson, Y. 221
Kolmogorov, A.N. 27, 100, 147
Koosis, P. 222
Kuipers, L. 223

Lebesgue, H. 39
Lebesgue constants 26, 30
Lévy, P. 185
Littlewood, J.E. 129, 132, 166
localization 7, 54
Loomis, L. 221
Lowdenslager, D. 121

Malliavin, P. 194
maximal functions 165
maximum principle 30, 44
McGehee, O.C. 221
Mercer's theorem 5, 129
Minkowski, H. 97
multiplier operation 61

Newman, D.J. 55
Niederreiter, H. 223

Ostrowski, A. 106
outer function 110, 123

Paley, R.E.A.C. 136, 142, 176
Parseval relation 4, 18, 24, 25, 84, 98
Plancherel theorem 62
Poisson kernel 28, 56, 58, 127, 135
Poisson summation formula 67, 69
positive definite sequence 40
positive definite function 69, 90
primary ideal theorem 195
Principle 15, 34
probability measure 74, 157, 167
probability space 171
product of groups 95, 100

Rademacher functions 170
Rajchman, A. 203
random variable 171
Rauzy, G. 216, 223
Riemann-Lebesgue lemma 8, 11
Riesz, F. 19, 41, 72, 117, 177
Riesz, M. 42, 152
Riesz, F. and M. 106, 107, 114,
    115, 133, 137, 141

Riesz-Fischer theorem	3	Tauberian theorem		35, 185, 190
Ross, K.A.	221	Theta function		69
Rudin, W.	204, 222	Thorin, G.O.		42
		Titchmarsh, E.C.		183, 184
Salem, R.	223	trigonometric polynomial		2, 83
Sarason, D.	110			
Schur, I.	130	unicity theorem		17, 54, 55
Schwartz, L.	191, 194	uniform distribution		205, 211
seminorm	101	uniquely ergodic		213
Siegel, C.L.	97, 223			
singular inner function	127	Vitali covering theorem		166
singular integral	157			
spectral set	191	Walsh system		172
spectral synthesis	194	Weierstrass theorem		18
spectral theorem	72	Weiss, G.		222
Stein, E.M.	222	Weyl, H.		205
Steinhaus, H.	102	Wiener, N.	22, 47, 64, 90, 111,	
Stone-Weierstrass theorem	84		136, 142, 181, 185	
sum of groups	95			
summability	30, 35	Young, Robert M.		222
symmetric difference of sets	96			
Szegő, G.	106, 118, 122, 199, 204	Zygmund, A.		43, 150, 154

Texts and Readings in Mathematics

1. R. B. Bapat: Linear Algebra and Linear Models (Second Edition)
2. Rajendra Bhatia: Fourier Series (Second Edition)
3. C. Musili: Representations of Finite Groups
4. H. Helson: Linear Algebra (Second Edition)
5. D. Sarason: Complex Function Theory (Second Edition)
6. M. G. Nadkarni: Basic Ergodic Theory (Second Edition)
7. H. Helson: Harmonic Analysis (Second Edition)
8. K. Chandrasekharan: A Course on Integration Theory
9. K. Chandrasekharan: A Course on Topological Groups
10. R. Bhatia (ed.): Analysis, Geometry and Probability
11. K. R. Davidson: C* – Algebras by Example
12. M. Bhattacharjee et al.: Notes on Infinite Permutation Groups
13. V. S. Sunder: Functional Analysis — Spectral Theory
14. V. S. Varadarajan: Algebra in Ancient and Modern Times
15. M. G. Nadkarni: Spectral Theory of Dynamical Systems
16. A. Borel: Semisimple Groups and Riemannian Symmetric Spaces
17. M. Marcolli: Seiberg – Witten Gauge Theory
18. A. Bottcher and S. M. Grudsky: Toeplitz Matrices, Asymptotic
    Linear Algebra and Functional Analysis
19. A. R. Rao and P. Bhimasankaram: Linear Algebra (Second Edition)
20. C. Musili: Algebraic Geometry for Beginners
21. A. R. Rajwade: Convex Polyhedra with Regularity Conditions
    and Hilbert's Third Problem
22. S. Kumaresan: A Course in Differential Geometry and Lie Groups
23. Stef Tijs: Introduction to Game Theory
24. B. Sury: The Congruence Subgroup Problem
25. R. Bhatia (ed.): Connected at Infinity
26. K. Mukherjea: Differential Calculus in Normed Linear Spaces
    (Second Edition)
27. Satya Deo: Algebraic Topology: A Primer (Corrected Reprint)
28. S. Kesavan: Nonlinear Functional Analysis: A First Course
29. S. Szabó: Topics in Factorization of Abelian Groups
30. S. Kumaresan and G. Santhanam: An Expedition to Geometry
31. D. Mumford: Lectures on Curves on an Algebraic Surface (Reprint)
32. J. W. Milnor and J. D. Stasheff: Characteristic Classes (Reprint)
33. K. R. Parthasarathy: Introduction to Probability and Measure
    (Corrected Reprint)
34. A. Mukherjee: Topics in Differential Topology
35. K. R. Parthasarathy: Mathematical Foundations of Quantum
    Mechanics
36. K. B. Athreya and S. N. Lahiri: Measure Theory
37. Terence Tao: Analysis I (Second Edition)
38. Terence Tao: Analysis II (Second Edition)

39. W. Decker and C. Lossen: Computing in Algebraic Geometry
40. A. Goswami and B. V. Rao: A Course in Applied Stochastic
      Processes
41. K. B. Athreya and S. N. Lahiri: Probability Theory
42. A. R. Rajwade and A. K. Bhandari: Surprises and Counterexamples
      in Real Function Theory
43. G. H. Golub and C. F. Van Loan: Matrix Computations (Reprint of the
      Third Edition)
44. Rajendra Bhatia: Positive Definite Matrices
45. K. R. Parthasarathy: Coding Theorems of Classical and Quantum
      Information Theory
46. C. S. Seshadri: Introduction to the Theory of Standard Monomials
47. Alain Connes and Matilde Marcolli: Noncommutative Geometry,
      Quantum Fields and Motives
48. Vivek S. Borkar: Stochastic Approximation: A Dynamical Systems
      Viewpoint
49. B. J. Venkatachala: Inequalities: An Approach Through Problems
50. Rajendra Bhatia: Notes on Functional Analysis
51. A. Clebsch (ed.): Jacobi's Lectures on Dynamics
      (Second Revised Edition)
52. S. Kesavan: Functional Analysis
53. V. Lakshmibai and Justin Brown: Flag Varieties: An Interplay of
      Geometry, Combinatorics, and Representation Theory
54. S. Ramasubramanian: Lectures on Insurance Models
55. Sebastian M. Cioaba and M. Ram Murty: A First Course in Graph Theory
      and Combinatorics
56. Bamdad R. Yahaghi: Iranian Mathematics Competitions, 1973-2007